For the privilege
of working together
to produce this.
Compliments Mario

[illegible]

THE SHAPING SPIRIT

THE SHAPING SPIRIT

JIDE ADEFOPE

NEW MILLENNIUM

First published in Great Britain in 2011 by New Millennium Press Ltd

New Millennium Press Ltd
64b Grange Road
Sutton SM2 6SN, Surrey

www.newmpress.com

A CIP catalogue copy for this book
is available from the British Library.

ISBN
978-0-9568613-0-6

Printed and bound in Germany

CONTENTS

	Preface	9
1	Will	17
2	Gene	57
3	Atom	74
4	Spirit	120
5	Time	214
	Epilogue	279
	Index	283

Preface

This book explores what it means to be human, what makes us who we are, and what it means to be you or me in spirit and intellect.

We share with other creatures an ability to experience the world around us. We are aware of never-changing laws that operate everywhere at all times; as human beings we are also dependent upon them. The laws of nature constitute our frame of reference in making sense of our circumstances. We seek to know and utilize these laws to our greater benefit. From the smallest to the largest natural phenomena, the unchanging, universal laws are uniform throughout time and space. Cosmologists, who study the birth and evolution of the universe, assert that our universe was created nearly fourteen billion years ago in a primordial explosion that is described as the 'Big Bang'. The same natural laws we recognize today have held true in the past and will be the same in the future. Experience of the laws does not modify them.

In the intellectual context of an age of reason, modern science is rising to a position of unquestioned authority. The historical quest to discover the nature of man – or humankind – has in the last few centuries been stripped of the certainties of old religious faith and observance, especially Christianity. Nowadays attention is more directed toward biological understandings. Interpersonal values and stable morality are, in general, glorified without reference to God. In modern society, men agree to live on the basis of secular rules. Having laid to rest belief in the arbitrary figure of a Creator, as suggested by the religions, does not imply leading an undignified, value-free life without moral reflection.

However, the acceleration of events at the present time presses toward a momentous enlightenment, a pursuit of truth and discovery that departs from atheism and the dubious conceptions of God.

There have always been 'disasters' on our planet, and every event is unique – though it may have parallels in history. This, however, is more than the realization, true of other ages, that we live in a world and a time riddled with chaos and suffering, with violence, hunger and deprivation, with abuses of power and injustice that occasionally swell beyond a civilized sense of balance, fairness and empathy. Our values are foundering. We are unwilling any longer to identify with institutions that have traditionally fulfilled our aspirations. Our environment too is degenerating. In an age of unprecedented affluence, nations sway on the edge of starvation. Wide areas of the globe suffer from plagues. Genocidal machinations are rife. For the first time in history we have the power to eliminate the entire life of the planet. Should we blame God, or the gods, or nature, or do we really only have ourselves to blame?

In this modern age of scepticism about the idea and concept of God, religious sensibility is challenged as a relic of unenlightened former times. We extol individual freedom above all else, together with the potency of the new gods of market ideologies and the engines of science that presume to give definitive answers to the big and small questions about existence. Human attitudes seem more tolerant under the principles of democracy, yet intolerance grows. Atheism spreads with dogmatic intensity, and with it the view of the universe as without purpose and of life as a fluke in the vast cosmic lottery of chemistry. Alongside this stand the fundamentalist caricatures of faith, with their footprint of atrocities.

There are notions that depict life and its development as being simply outcrops of blind chance, the earth no more than an insignificant speck in the universe, and free will only an illusion of the brain. In mirroring social trends, art clearly sets forth this acceptance of an existence in which meaningful development is of little account. These views jeopardize humankind's most basic inner hold, namely, that everything we will, think and do is meaningful and can make sense.

The bewildering presumptions about the human condition leave an interminable inner emptiness, a gnawing self-doubt shrouded in a moral fog. Nonetheless the questions arise: if the world around us is the work of an all-good, all-powerful, omniscient intelligence, why then is there so much apparently pointless suffering and evil? How is the existence of evil compatible with the existence of a benevolent God? How can God allow such things? Why do good people meet an adverse fate? Are all events predetermined or do we have free will? Are we who we are because of free choices we make? These questions are a testament to twenty-first-century tolerance – in former times they were blasphemous.

Our concept of freedom is linked with a sense of selfhood, and feeling free to make choices is at the heart of our sense of self. Our self-consciousness grows in the immediate certainty of its recognitions. We are sufficiently self-aware to realize our limitations and know that free choice can have its constraints:

- The self can be free to the extent that its freedom does not infringe on or deprive other selves of their own freedom.

- Social harmony rests on the attainment of a degree of personal self-freedom.

- One person's unrestrained free choice may run into conflict with another's.

- We may be limited by our former decisions, our biological nature, or through tailoring our options to the expectations of others.

We need limits to our capabilities to enable us function and develop properly. We cannot fly like a bird or swim like a fish. This would be contrary to our constitution. We cannot live in deserts without water or in frozen wastes without warm clothing. We are also in peril from storm, tempest, earthquake and fire. The predicament of natural and man-made restrictions on our freedom encourages our shared responsibility, one for another.

Whatever place we occupy in a determinate reality we are privileged with the ability to think. We are motivated in our conscious state of vigilance to reconcile mind and body, to achieve a coherence between our inner and outer life.

Every person has an inner voice that resonates in our self-directing motivation, in the inner yearning for self-improvement, and the ideal of personal perfection. We understand the otherness of others, or that others have minds; we may gain some insight into the minds of others and even enter deeply into their thoughts. We may not therefore be overly self-absorbed in interpersonal relationships. As obscure as the motivations for other people's behaviour may be, it is nonetheless possible for us to relate to them in love, empathy and compassion.

Unlike other creatures, we can dissociate ourselves from situations through our capacity for altered states of consciousness. We can retreat inwardly from terrible external surroundings to an inner life of freedom. We humans are inwardly awake in a way words cannot describe. Although we are subject, like everything else, to the forces of nature, we feel we have a duty to truth, a simple intuition for what is good and what is evil. Well, some may; others may not. Nevertheless, moral reflection marked the beginning of human self-consciousness. The freedom to make choices implies inescapable responsibility – to express the full force of our developed personality in expanded awareness. The energy of life is a motivation to engage the world in experience and to realize our inner capacities by manifesting them.

How do we explain the fact that our universe is as it is? Scientific innovations over more than a century have brought greater understanding of the natural world, from the microcosm of atoms and elementary particles to the macrocosmic universe that these particles make up, in areas such as cosmology, genetics, neuroscience, and quantum mechanics. However, intuitive perception, our capacity for immediate insight and certainty unmediated by observation or rational analysis, hints at the mastermind of our conscious experience beyond the brain. It suggests more profound truths than the fragmented categories of piecemeal knowledge that can be derived from rational analysis after the fact by present-day science.

As far as we know, no other creature has developed the capacity of thought to the extent that we humans have. We actively plumb the deepest mysteries of our universe. We want to know who we are by understanding the world we live in. What living power is at work in the all-pervading coherence of the cosmos? Not a scintilla of the great balance can be altered without plunging everything into ruin. When was the Big Bang – if there was one? When was time born?

Historically, our education on the cosmos has been left by scientists and the culture at large to philosophers and theologians. Today, many primary religious and philosophical questions are contemplated through a manner of thinking called the scientific method. It is a quest to fathom the use of the cosmic laws for the greater benefit of humankind. One of the main principles of science is that the natural laws apply equally across all scales of reference. The scientist takes carefully observed natural phenomena and seeks to discover and formulate the framework of a theory to unify its various aspects in a revelation of their unrecognized relationships. The scientific method uses formal rationality in the making of argument or discovery of fact. Science queries why the natural world is the way it is, premised on the notion that the human mind can comprehend and construct an intelligible rational explanation of the vast scope of cosmic phenomena. But the more insight we gain through science into how the world works, the more we are reminded of our relative insignificance and humbler role before the vastness of the cosmos.

Einstein's theory of relativity shows us that our world is merely a speck of dust in a unified geometry of time and space strewn with comets and numberless stars. These stars are so far away that their light takes millions and billions of years to traverse seemingly boundless expanses of space to reveal information about their conditions to us here on earth. Peering deeply into space, spanning as far as the telescope can reach, is tantamount to looking far back in time. What we see on the brightest summer's day or across the glittering canopy strewn over the clearest moonlit night sky is but a shadow of what is really there. We see stars that are no longer in existence. There is more to seeing than meets the eye. We cannot be certain of what we are seeing, or even that it is there in the first place. Seeing is *less* than believing. We literally cannot believe of the world what we resolve with our eyes. Things may appear one way and yet truly be very different. What we think of as real is not necessarily so. Our view of the world changes as new information emerges, and as ideas are clarified.

Modern microelectronics, with microchips and computers, radio- and telecommunications, is the result of the bizarre paradoxes of quantum mechanics – for a century an outstanding exposition of the fundamental aspects of reality. Quantum mechanics correlates the particle and wave characteristics of light and matter. The paradoxical rules that describe the behaviour of matter and the energetic forces swirling in the cosmos form the foundation of much of the modern world. Quantum mechanics explains the stability of atoms and has given rise to technologies that have changed our everyday lives, from cell-phones to lasers, transistors, CDs, microwaves, x-ray machines, astronomical telescopes and communication satellites. It explains chemistry *and* physics, and underscores the stability of the DNA (deoxyribonucleic acid) molecule and thereby the functioning of the living body.

We live in a time of unprecedented expansion of the frontiers of knowledge. The knowledge explosion of the last century or so has not only burst outwards but inwards as well. In the wake of Freud, the Darwin–Wallace selectionist vision of evolution, and Watson and Crick's discovery of the structure of DNA, we are no longer masters in our own house. Whatever claims we make about being in charge of our destinies, most of our behaviour and mental life is stirred by drives that we are barely conscious of. We process most information mentally. Most of our everyday thinking, feeling and acting occurs outside our conscious awareness. Many things, like breathing, speaking and walking, or typing, playing a keyboard or driving are done without conscious intent, as though we are sailing through life for the most part on autopilot. From childhood on we speak, uttering strings of words in coherent sentences, even at times absent-mindedly, with little or no idea of how we come up with these words. Consider skilled guitarists or pianists, who can carry on a conversation while their fingers play a tune.

Psychologists describe humans as creatures whose actions are predetermined by impulses, with proclivities to pleasure and power, that are rooted in early childhood experiences and social conditioning, and whose life progression derives from their genetic inheritance and the workings of the brain. In their genetic makeup, children reveal the physical characteristics of their parents. Genetics demonstrates that there are only slight differences between the genetic makeup of humans and apes, and how the genes of humans and several other species correspond closely. The relation of genes to personality and behaviour remains a burning issue, and draws contrasting arguments. Meanwhile, the wealth of genetic data now at the disposal of industry, medicine and agriculture holds major implications with regard to personal freedom and individual conduct within the community.

As our knowledge of the world without and within expands, so our sense of self, together with the emotional satisfaction we experience concerning the sense of our own importance, diminishes. What makes us human, or, for that matter, inhuman? Each one of us has developed physically from our beginnings. Is there such a thing as the *core* of the self, the centrality that develops in awareness of itself and distinguishes us as individuals, and, moreover, as humans from animals? What is the place of consciousness in a material world? How does *will* appear in consciousness? Does consciousness emerge totally from the organ of the brain? How do mental and physical processes enable conscious experience, with its subjective qualities? What is the relationship between mind and the neural components of the brain? How is the active energy of the spirit expressed in our bodies? What is the experience of spirituality in a secular world? Or are we simply wandering groups of cells, with minds that are merely a passive reflector of sensations and passions, set against a backdrop of nature, with no particular threshold that determines human individuality? Thus, are we pressed in drive, impulse and desire of experience by the cascade of neurons in the brain towards self-realization? The debate over whether we are different from animals has gone on since ancient times, and has occupied the minds of philosophers and biologists for generations. Today we discern clues to our origins and behaviour among the great apes. Could it be that, despite our proud boasts about our mastery of the natural world, expressed in the imagination of artists and the experiments of scientists, we are simply beasts with sophisticated brains? We like to think otherwise, that we are set apart by our spirits, the wellspring of our awakening in conscience to a profound imperative of moral personality. Spirit impels our moral aspirations and motivates us to higher culture. At the very least, we can point to a wilderness of monkeys and show the difference.

ONE

WILL

Never have so many people been exposed to so much knowledge. We have lived for millions of years on this planet but our systematic eye on the world has developed only in the last ten thousand. Our modern beliefs about how the cosmos began are based on the premise of a creation event out of a void, that our universe emerged from a primordial 'big bang' fireball close to fourteen billion years ago. At birth, it was in a condensed, hot state without structured matter; it cooled as its volume increased to give rise to *all* matter, energy and space, and the universe has been expanding ever since. The origin of the nascent mass of super-dense radiation remains controversial even after more than a century of uncertain philosophical and scientific debate: is it a random fluctuation in the finite past with cosmological consequences? Or is it a controlled creation of time and space? Chance or purpose? We are not sure which. Within a blink of the primordial explosion, in that first split second, the building blocks of matter and the forces that structure the universe were laid down. Under unbelievably great pressure, the thermal soup of fundamental particles cooled down, so that the expanding universe began to coalesce into subatomic particles, atomic nuclei, atoms, elements, compounds and cosmic bodies. Utterly novel forms emerged on new levels of reality, different levels of organization and sets of arrangements evolved into the world that we see around us – stars, galaxies, and the teeming planets, at least one of which, the earth, has been able to generate life and support conscious beings, such as ourselves. Was the big bang the beginning of the universe? Or are other realities of space and time possible in a historical timelessness?

Rhyme and reason

Inexorably, life is spawned by the natural environment. The world around us abounds in changes, on both small and large scales. To survive, living organisms must adapt to their environments. Manifold adaptations are apparent everywhere in the immense variety of the natural world. Human beings share with other creatures an ability to respond to their surroundings, and must adapt inwardly and outwardly, passing on genetic information from parent to offspring.

As humans, the body of flesh and blood, together with our inner spirit or self-conscious personality, makes us who we are. We are aware of our own self, of other selves and of the self-evident world around us, and we are conscious of our individual experiences. We recall our past, explain our present, and predict our future, and we can model alternative futures for ourselves. Genes determine the nature of our body. They dispose us to certain abnormalities and protect us from others.

Intimations of immortality are on the wane in the belief that the material world is the only reality there is and that we fulfil all our aspirations within it. Scientists seek to unravel consciousness in the chemical and electrical actions of the vast jungle of nerve cells inside the human skull. Yet the brain is discovered to be a system that organizes itself without a special command centre, and consciousness has no certain place in the brain. The most basic function of the brain is to keep the body active. If there is such a thing as free will, it would imply a relationship of our personhood that transcends, and is independent of, any determinate mechanism of the universe, including the brain.

We characterize the natural world partly from the way we think and record experience. Scientific descriptions of the world fascinate the popular culture as enduring fads. Some imagine that only science promises the most comprehensive and predictable explanation of natural phenomena. Science, it is believed, provides sovereign access to the discovery of truth. Biological evolution, popularized in the mid-nineteenth century, which links all living things by common ancestry in a uniformity of nature, has almost attained oracular status in our time. The gene, or even something smaller, seems to have replaced an omnipotent Creator in our understanding of the nature of life and existence. Predictable laws, the 'truths' of science, explain human behaviour, the behaviour of atoms and of the large-scale universe. Life, transformed from simplest states to an extraordinarily complicated creature, is viewed as having no meaningful values outside biological adaptations. We are left to create our own values without a free conviction or even being able to know that such values are right. Science, which is all about what can be weighed and measured and what our senses, assisted by sophisticated instruments, can correlate from the world of phenomena, cannot provide a foundation on which to judge the morality of any values we come up with. It has no time for a God with humanized qualities moving in inscrutable ways to intervene outside the lawful mechanism of the universe. Life is formulated as the concerted expression of cell activity. The human body consists of roughly sixty trillion cells, and each cell has about ten thousand as many molecules as the number of stars in the Milky Way galaxy. The cell is the basic building block of living things. It is sustained in a particular molecular organization that obeys physical laws irrespective of our intentions. Genes equip our constitution and affect our behaviour. For 'modern man' it is not the infinite wisdom of the Almighty but the enigmatic little gene that is lord of the living and the dead, that establishes arbitrary boundaries between the living and non-living.

Every human being is unique. No two persons are identical. Yet in a world of roughly seven billion people, each is naturally similar to every other person. We may look different from our neighbours or from people at remote distances from where we live, but we are alike genetically. We also share numerous traits with the rest of life on the planet. Many genes are common to plants, humans and animals. All living things have pervasive affinities and have evolved from forms that existed before. Animals communicate with each other by gestures or sounds, but only the human mind has developed the complex system of symbols we express in our language to transport thought into the mind of another and to open our hearts to other human beings. Bees can 'talk' by dancing on the honeycomb or in the air, but they can only talk about food sources or other immediate events. We can talk about *anything*.

We need to ask questions and find answers, and in the process expand our consciousness. The deeper our conscious awareness, the more alive we are. Unlike the monkeys, with whom we have a genetic kinship, we seek not only food for our material survival but also food for thought. We look backwards and forwards, inside ourselves and outside. We have arrived at truths and come up with disciplines describing workings of the natural world that can be tested, and which test our capabilities.

Genes are carriers of hereditary information about the physical constitution that pass from parent to child. Like begets like, and children often resemble their parents. Genes embrace chemical elements of encoded information, the integrity of which cannot be compromised. They withstand the test of time, unless alterations are introduced as an adaptation to long-term changes in the habitation environment. Genetics explores mechanisms of inheritance and correlates changes in DNA with changes in how a person, animal or plant looks or behaves. It describes how we become who we are, and may hint at how best to change who we are. Our future is no longer written in the stars, but in our DNA.

The stars are still there though, sextillions upon sextillions of them, like tinsel lights glittering in hazy swarms of galaxies and comets trillions of light-years across the firmament and, like our sun, often with a company of satellites. How long have they been there? How did they get there? How long will 'our' universe last? How will the world end? Or will it? Today we are not persecuted for asking such questions – we have to ask and are encouraged to ask. Our future depends upon asking. It is the essence of our being to ask. We want to know where the massive unseen energy that pervades the farthest reaches of the cosmos, shapes the planets, suns, stars and galaxies actually comes from. The universe is expanding, and so is our knowledge. But how did we get from void to the universe? Where do we come from? What basic stuff is the whole world made of? To find out, we look to larger, more sophisticated and accurate instruments.

Moving parts

Research into the nature of reality on the subatomic scales of time and distance, in the form of quantum mechanics, suggests that matter is perhaps more illusion than reality. The world of natural phenomena no longer appears completely deterministic but is realized in a multi-dimensional matrix of probabilities. The Heisenberg uncertainty principle, a cornerstone of quantum mechanics, states that the exact location and precise velocity of a particle such as an electron in a small region of space cannot be determined at the same time; and conversely, measuring the precise velocity of a particle makes its exact position uncertain. In other words, we ultimately determine our own reality through our actions and perceptions. This concept of reality defies our intuitive common sense of time and space; yet quantum theory has allowed researchers to calculate processes of microphysics with fantastic precision to produce modern computers and electronics, lasers, nuclear reactors, nuclear magnetic resonance imaging machines and applications of radioactivity in medical diagnosis and therapy. And the atom bomb (although it is hardly an achievement we can be proud of). In outer space, different orders of combination of subatomic particles are, as we have discovered, finely balanced in the nuclear furnace that glows to make our sun shine, dispense warmth and light, and make our planet evolve.

Quantum theory suggests an abstract foundation of energy and its relationship with matter. The dilemma is that having discovered – some would say constructed – a rational order behind this wondrously complex universe, we can find no *purpose* for its existence. The more comprehensible the natural world gets, the more aimless and pointless it seems. The generative forces of nature appear to transcend matter. The laws seem ultimately arbitrary and have no reason to exist. At the same time, processes of subatomic particles make up the properties and behaviour of the large-scale environment, with all the implications of the paradoxical quantum entanglement at the heart of matter. All at once, faced with the realization that the wisdom that governs the perfect engagement of the cosmic laws is beyond the human mind to presume to index – let alone harbour – optimism in our ability to come to grips with the intricacies of the world is waning. The patchy descriptions cradle life in blind chance amid a pervasive randomness in the physical world. Once near to angels, we are brought down low to contingent creatures in a contingent cosmos. What about the values that make us human, such as beauty, love, joy, loyalty, justice and other deep perceptions – all that we have ever nurtured as precious and which are not subject to analytic understanding? Are they mere illusion or do they reside in a domain outside the framework of science?

The idea that species appear by selection in evolution, conceived a century and a half ago by naturalists Charles Darwin and Alfred Russel Wallace, instituted a biological understanding of humankind and implied a common ancestry of living creatures. Conditions on the planet more than three and a half billion years ago opened the way to quite simple life forms. Darwinism – as the theory of selection in evolution came to be known – posited that all of humankind is to be seen as on the same continuum as the animal, with no essential distinction between the two. Life is a struggle in the mad chaos of nature for survival, in which populations compete for resources, for food, water, territory, and opportunities to mate. If breeding results in more offspring than the resources of the habitation can support, the struggle to survive keeps populations at roughly the same size, with modest fluctuations. Changes occur in the gene pool and incremental modifications may lead over time to a new level of form, perhaps even resulting in a new species.

Since Darwin, the imperative of struggle has become raised in the cultural and political consciousness, as well as verging on a moral code of society. Strong genes will survive; the weak inevitably face extinction. And rather than duplicating like they do in bacteria, some genes engage in close-fitting relations in the fruit-fly and the elephant and in a really complicated and intelligent vector such as a human. They breed to fashion more diversity. Darwinism turns the human into a temperamental supercluster of cells, a survival machine, a robotic happenstance of transmission of heritable variations from generation to generation in the sweepstake of births, deaths and relationships. We are animals roaming around with no greater purpose than enhancing our genetic potential. It paints a grim picture of the human condition, studded with spectacles of the weak having to fall by the wayside. That is nature as it is, casual to disaster; we cannot all survive and we will all eventually fall by the wayside.

It must occur to thinking people to question why organisms more complex than bacteria should exist. The probability is that we will never succeed in developing antibiotics faster than bacteria can adapt to them. Drugs wipe them away by the billion, but the few survivors introduce a newly evolved biological strain. The self-awareness of humans, at least, appears to set them apart in an insuperable manner, and is needed to make meaning in everyday life. Man, not the bug, is the maker of meaning. Thus the tireless struggle for improvement pertains to his inner nature.

Science and art, cultural vehicles in which our interpretations of the timeless order of the natural world are recorded and realized, and with which we can engage and define ourselves as individuals, are feats far beyond the reach of bug and ape. Where is the human being who derives no meaning from the realms of physics and mathematics, of theology, aesthetics and philosophy? We are, within obvious limits, free to choose our own path.

On the trails of ...

We know we have a free will. Willing consciously is, for us humans, an everyday experience. Of course, we also belong to an external objective reality; the expressive world, governed by cause and effect, limits us from exercising absolute freedom. We are subject, like all creatures, to natural disasters, to ambush by predators and to the constraints of our biological constitutions. But we can, to a certain extent, surmount instinctive behaviour and live in control of our passions. We have the capacity in thought and action to will our own ends and purpose, setting us free of the constraints of physics, our genes, our history and even time. The affirmation of free will is the bedrock of the rule of law, the rights and personal responsibility on which civilized society stands. This responsibility gives rise to the moral dimension and the convention of laws that we enact for our social arrangement and civic harmony. Balancing personal freedom with a sense of collective purpose is at the heart of our moral dilemma. Whether or not our concepts of good and evil are arbitrary, it seems a fallacy to infer morality from the laws governing the natural world. Is there a moral law of the psyche analogous to the unchanging physical laws? In an uncertain world, might a moral voice be counselled by conscience toward moral behaviour? Or are moral norms and values merely etiquettes rationalized under instinctive and emotional strains to peer or social pressure? Choice dominates our life. To what extent do we do the right things by conscious choice, or because of instinct? Rational enquiry provides no guidance for the ethical and moral issues humanity grapples with. Is there such a thing as justice? By definition it is a good thing. But how, in a world of manifest unfairness, is justice naturally to be attained?

Choice affirms our self-motivation and moral responsibility. Freedom of action is necessary for moral responsibility. Without an acknowledgement of free will, without the freedom to choose, we cannot have values and nobody could be held responsible for his or her own actions. We are not determined by our biology. Should we excuse tyrants, megalomaniacs, genocidal dictators, torturers and terrorists, or accept as natural whatever child molesters take it in their heads to do on the grounds that they are only acting in accordance with the emotions and desires resulting from their inclinations and predispositions? We can question our own motivations. Without free will, our ethical values and aspirations fly out of the window. We consider justice is done when good is rewarded and evil is penalized. The principle of cause and effect, which is self-evident in the world around us, grows in our awareness from early childhood. However reluctantly, a misfortune we know to be just intuitively unnerves us less than one we cannot explain. Unbidden fortune we call *luck*, even if at the same time we assume that we deserve it.

The free exercise of our will is not founded on something other than ourselves. It is clear to us that every proper decision originates from our sense of self, from our inner being, and not just from our body, or from our brain. We know ourselves from the *inside out*, not the reverse. The brain does not generate the self, although, unfortunately, many people presume the brain is the centre of consciousness, rather than just another organ, albeit an important foundry of thought. This goes so far that the cessation of brain activity, or more precisely brainstem activity, is used to ascertain death. Sleep and wakefulness define our primary states of being. When we fall asleep we lose consciousness and we regain it when we awake. The wakefulness of day-consciousness is dependent on brain activity and is linked with the ability to think.

Dream experience, in contrast, occurs when the waking state is turned off, for example in sleep, and our self-consciousness is loosened from our physical body. Dreaming and the waking state are unblended realities and dreaming is, a lot of the time, a far more intense perception than the thoughts that occur in day-consciousness. Self-awareness transcends brain events. The landscape of our inner being, that is the nature of the self, is vastly more profound than the perception spurred by the bodily senses, and it alone is the basis of free will.

Free decision of will is primarily not an attribute of the brain. It belongs to the 'I', the spiritual ego or self. Free will lies in the spirit's vibrations of intuitive recognition, a unique cognitive attribute of immediate certainty that feeds the hunches, gut reactions and snap judgements we apply nearly instantly, even before we can shape words for events and circumstances, and in many ways shape the course of our life.

The human mind today is, to some extent, seen through the separation made between the otherworldly and the earthly in explaining the relationship of the soul to the body, as with the body–soul dualism of the mid-seventeenth century. René Descartes constructed a mechanical model for all living processes, suggesting that consciousness exists independently of the brain, and can act on it. *Mind*, subsequently narrowed to a psychological entity rooted in empirical interactions of brain cells without a relationship of transcendence, came to replace *soul* in a secular dualism, and eventually led to an understanding of reason as the cardinal human attribute. The soulless personhood was born.

However, both the physical world and the so-called beyond or ethereal hereafter are in reality united by the fundamental natural laws to which all processes are subject, be they spiritual or scientific. Spirit seeds self-consciousness and human identity. The cloak of the physical body, in its details determined by genes, enables the spirit spark to manifest in the here and now. Expanding self-consciousness is an expression of the spirit unfolding and human development. Spirit in this world is outside its sphere of origin. The body belongs to this world of coarse consistency. Personal identity unfolds through motivating the mental state by the spirit's volition. Volition expresses itself through our embodied thinking, so that not all thoughts have a physical cause nor do they originate in the body. The disembodied spirit stands outside the influence of the body's intellectual will.

Voices reiterate soul and self-realization, yet there is brazen pursuit of material self-interest. Everything is geared to stimulating and satisfying physical appetites. The craving to satisfy the body has eclipsed aspirations of the spirit, its creative whispers muffled by our subordination to the intellect. Religious anthropology, addressing a redemptive capacity to transcend our animal nature, has been supplanted by secular anthropology out of a desperate passion for a better world.

In earlier times we found solace in the clouds of religion, for 'heaven' simply meant sky, and Heaven held the antidote to the manifest inequity of the here and now, and revealed all. But when Heaven, together with its assurance about the afterlife, went out of fashion, we turned to the mundane world of Science, the atoms of Democritus, the compasses of Newton in optimistic ventures of self-enlightenment. Science, with its empirical rules of argument based on experiments and observations, has nothing to say about otherworldly engagement or divine action within nature. For many, human nature is to be equated with physical nature, and the model of a phenomenal universe devoid of otherworldly connections and restricted to verifiable facts, holds the key to cosmic truth. Today, amid the claims that modern science is the paradigm of all knowledge, there are many who think they can do without God.

The paradox is, the more we discover, the less we know. Each giant step we take diminishes us. We have put our faith in science and it has provided answers that create more and more unanswerable questions. This, of course, does not mean that we should shun science, but we must come to the realization that it has its limits and that we must engage all our abilities to the fullest extent in order to be able to carry our investigations further.

Science seeks to explain reality through weighing and measuring and by conducting reproducible experiments. In probing from below, using mostly the intellect with its framework limited to earthly concepts of space and time, only calculable aspects of a small-scale context of behaviour are reckoned with. Countless factors in the large-scale engagement, often unknown ones that play a big part in decreasing, strengthening, or altering the effects, are ignored or rejected, so that the final reality can be the exact opposite of what the best science can claim. For example, the scope of reality suggested by quantum mechanics is out of step with the finite metaphysical resources commanded by the intellect. That is why science so far cannot reconcile its descriptions of gravity across the rules that govern fundamental particles and those that govern big objects that the particles make up, such as trees, buildings, cosmic bodies, galaxies and us.

The intellect is a product of the cerebral cortex that constrains our recognition within a context of evidence-based analysis. Overdeveloped and over-cultivated, it suppresses the far wider vista of intuitive recognition and limits us to earth-bound conceptions of space and time. It is as if the moon were to think that with the solar eclipse it had quelled the blazing sun. The erosion of our spiritual capacities is a negative anthropology with no way out of the ensuing cultural despair. Reliance on brain power without immediate certainty of intuitive direction is at the root of what makes us *in*human.

Our species has wandered the world for millions of years. But what am *I* doing here? We imagine that what we see, touch and feel with our earthly senses constitutes the reality of the entire Creation. We saunter along indifferently through life, with only small earthly aspirations and goals in mind. The sacred duty to investigate the real purpose of our existence hardly touches us. If it does, we tend to think there is no answer.

Pride comes before a fall. This adage pervaded the thinking of medieval people and the Greek tragedians. Man is a creature. He is not king of Creation and he has no power over nature's law. He can only act with a freedom of volition in an eternal order beyond grasp of the intellect, neither with mathematics nor with science.

As there are cycles in nature, so there are also in humanity's thinking. It is sobering to think that for all their (to us) marvellous advances, science and technology have even given us the tools to destroy ourselves. The rush for faster and more powerful technology brings no depth of inner life – quite the opposite, in fact! For those who know what to look for, all the classic signs of a civilization in its death throes are evident in every sphere of public and private striving. The same disintegration was the hallmark of the last days of Greece, Rome and Byzantium. To be sure, the world has been convulsed by upheavals throughout history, and it could be argued that the ever-more intense disasters are perhaps no more than a natural variation. However, the present time presages an unmatched pace of disintegration and collapse, turning our subtlest scientific weapons against ourselves and posing a real threat to our very existence on this planet. The intellect is incapable of extricating us from the vicious circle of fear and distrust simply because humanity's total reliance on this intellect at the expense of the spirit is the primary reason for this sorry state of affairs developing in the first place. Many can sense in the painful changes and erosion of values the transition to a new spiritual adjustment, without which humankind will have no future at all. The decline and disintegration ushers in new attitudes and new perspectives on reality beyond the purely material metaphor, not limited by science, religion or any philosophy.

Fervour of being

We are on one level physical beings, embodying thoughts, feelings and a lot of what happens to us in our mental and social experience. The same organic components are common to all living things. On another level we are far more than our physical nature. We desire and perceive, we make decisions in everyday life, and we feel emotions such as joy and sorrow. We acquire an original and bountiful source of experiences that awakens a deep awareness within us, upon which we can draw to make decisions or determine the general direction of our life. A child grows in awareness of its surroundings and into a personality that is aware of an inner self. The sense of self begins with the child's sense of the world. He comes to perceive that he has a body that is separate from his sense of self. It is not simply that he *is* a body; he *has* a body. The 'he' is different from the body. Within obvious constraints, he can do what he likes with his body. He has an ability to choose.

We have affinities with other forms of life on the planet and yet we are unique. We have self-awareness and can dissociate our self from our body through the power of thought. Unlike animals, with their impulses and inclinations, intuition shapes our superior recognition and gives us freedom in our decision making. We seem born not merely to survive and leave descendants, but to experience the natural order, to examine the deepest aspects of the self and of the world at large.

A distinctive human feature is an enlarged brain. Rhetoric apart, no research into how the brain works can tell us the nature of consciousness, how the body becomes mind, or how the grey mass under our skull can elaborate our sense of self. The brains of many creatures simply respond to stimuli, however inventively, without the authority of self-awareness. Brain impulses and nerve electric potentials, together with conscious processes, may be part and parcel of the same activity, but intuitive self-awareness is a distinct capability that transcends the biological determinism of the brain. It feeds our purpose for self-expression and for answering for ourselves the deeply consuming questions: Why and how have I come into being? Why am I here? Where did I come from? What is my purpose, my goal? What is the meaning of my existence?

Fresh from nature's mould

Religious and scientific narratives posit polarized models of the cosmos. Judaeo Christian anthropology, for instance, installs man as the crown of Creation. The world is created for his benefit. He is a surrogate lord, entrusted with the natural world. The sun is put in place to warm him, and the stars to guide him. In contrast, the science of evolution dethrones man to an animal like any other, spawned from a common ancestor with the great apes amid the contingencies of the driving force of nature that is without essential hierarchies from God to beast to man. Human existence, in short, has no more cosmic significance than that of a bug or stone. Genetic scientists attribute life to encoded chemical letters of the DNA, where humans are an animated cluster of cells with a capacity to communicate, plan, make a vast array of tools, adapt to and modify a wide range of environments, and engage in cultural diversification.

The conundrum of existence is at the root of philosophy and the quest of science for universal explanations, and it is behind religious ambitions. A great gulf separates our enlightenment from the incapacity for reflection that is typical of other species, even though we cannot compete physically with many animals. Instinctive behaviour is an inexact guide to reflection of the great complexity of human consciousness. Persistent change, as animals pass on characteristics to their descendants over generations, is not to be seen on the same continuum as our sense of self. But new species do develop in immense creative cycles over huge tracts of time and we do indeed have much in common with other mammals.

Humans have evolved over time and are genetically related to African apes, most closely to chimpanzees. The hominids, an ancient family of upright-walking primates, spawned the lineages of African apes and the earliest humans. Hominids embrace an early stage of human evolution. Early hominids resembled and behaved like apes, but were different in that they could walk upright on the ground. This innovation is in contrast to the wrist-knuckle walking of chimpanzees or gorillas. Some six or more million years ago in Africa we shared a common ancestor with the apes, when the split occurred between anthropoid apes and human ancestors, close to the time when the ancestors of humans and chimpanzees parted company. It is not that modern humans have evolved from living chimpanzees; both descended from last common ancestors that were neither chimpanzee nor human. From the ape-like ancestors, a hominid lineage led to us, with our anatomical modernity; and divergent evolutionary pathways led to living African great apes. Nor did an ape-like hominid one day produce an offspring that looked like a human of the kind that lives today. Our modern form spans arcs of cultural transformation and historical change, in parallel with changes in DNA sequences and genetic expression removed from the apes. Chimps, too, have continued to evolve since their evolutionary split from human ancestors.

Except for modern humans, the hominid family became extinct. The last hominids that coexisted with our species, but lived in different geographical regions, were the larger brained Neanderthals. They were competitively displaced as a result of the expansion and the dispersal of early modern humans from Africa and subsequently they died out. Our species is the epitome of hominid evolution, but this raises the question whether australopithecine (an extinct species of small-brained hairy apes already wandering upright adeptly on two legs in Africa four million years ago and ventured from the rainforest into woodlands and savannahs) was essentially like us, with consciousness of aim or purpose, imagination, free will and a sense of responsibility.

A moral awareness separates us from the beast, with our sense of right and wrong, virtue and vice, and the codes of appropriate behaviours. We have an open choice of action; the beast does not. Action for the beast is less problematic, for its instincts tell it how to act, rather than its conscience. The animal acts in accordance with its impulses or inclinations and on the basis of what it has learned. Conscience – literally the knowledge that is within – assigns morality to our behaviour. It alerts us to do what is right and to will what is good. Our other qualities include love, joy, creativity, a sense of beauty, truth and self-respect – and, in turn, also their converses.

Naturally enough these higher attributes are species specific. A gazelle does not know it is elegant, nor does a vulture feel cruel or rapacious. We may feel it a shame that roses have neither eyes nor noses, but that is the way they are. If a tabby cat has a sense of beauty, we know nothing of it. And is the dog's submissiveness to people merely behaviour honed by thousands of years of content interrelationship? We may attribute moral values to animals, but they are in most cases certain to be human *projections*. A creature by its nature cannot exceed the boundaries of its species. We speak, but apes do not. We laugh, but the joy-crazed dog does not.

Footprints on the shores of the unknown

Fossil and genetic evidence locates human ancestry in Africa. From time to time the fossilized bones of an early human species, unearthed in the plains of the African savannah, excite a flurry of speculation. Human fossil skeletons from the first two to three million years of our ancestry belong entirely in Africa. The earliest presence outside Africa was much later. There is no identifiable fossil evidence among these African skeletons, of an ape with uniquely human features that is distinctive enough to be represented as a direct ancestor of modern humans. Primates are a category of ancestral mammals, including humans, monkeys and apes that share many characteristics, such as greater stereoscopic vision, the fine ability of the prehensile hand to grasp and manipulate objects, and a relatively large brain. Apes come closest to humans in their brain and body form, and share some human social and cultural attributes.

Humans show close affinity with primates like gorillas and chimpanzees. Like some other hominids, they are remarkable for their ability to walk on two legs, their slow rate of maturation, large brain size, and their toolmaking and tool-usage. Hominids and early humans display adaptations of standing and walking upright, the expression of language, culture and social organization. From the humble beginnings of our ancestors, our mental transition has enabled us to surmount the vagaries of our natural environment and become the dominant intelligence on the planet.

Transitional fossil specimens, with a mixture of ancestor and descendant features, prevail *within* species but are thin on the ground *between* closely allied species. The patterns in the fossil record show both a slow and steady, as well as a rapid and periodic, pace of evolutionary change from ancestor to descendant. There are many lacunae in fossil records, which suggest that every few million years a new species makes a sudden appearance in the biosphere, with little transition evident between ancestor and descendant. The hominid fossil record, documenting millions of years of human and ape evolution, is far from complete, with evidence that is often fragmentary, and the entwined description and interpretation of such finds generates unending speculation.

Primates are thought to have started evolving fifty-five million years ago, and the early anthropoids appeared over the following twenty-five million years. Fossil evidence for lineages of monkeys and ancestral apes has been dated back to Africa twenty million years ago. The jaw fossils of apes from around ten million years ago have affinities with those of early humans. Fossil findings reveal bipedal apes with big brains living alongside the australopithecines between two and five million years ago. There is, so far, no evidence of the existence of an upright biped who is part human, part ape, with a face that is smaller than the ape's and more vertical than snouty, and feet designed for standing rather than climbing.

Animals and plants change through time as they struggle to adapt in the dynamic conditions of their environment and propagate their species. No two individuals in a population are completely alike. Evolution selects for reproductive success and is consistent with passing on that survival advantage to descendants. Favourable survival characteristics become accentuated across generations at the expense of less well-suited traits.

Populations that reproduce, producing more offspring than can possibly survive, will in the struggle for existence kill both their natural prey and their own kind. In conditions of scarce resources, any number of organisms will not survive. So, too, the fate of the dinosaurs, once rulers of the earth, demonstrates that past reproductive success is no guarantee for continued survival. Changes in the environment will press some populations to grow legs and lungs, some fins, and others feathers and wings. New shapes may appear or the proportions and relations of body parts may be altered. Species and populations may die and others replace them.

The gene is at the heart of adaptation. Like breeds like, with changes that are introduced seamlessly through the generations so that the descendant may differ markedly from its distant ancestor. New foods need to be found, and new territories colonised; otherwise a population will die out.

Liberty of heritage

Early man, *Homo sapiens* or 'wise man', appeared on the continent of Africa at the beginning of the Pleistocene glacial period around two million years ago. *Homo sapiens* was the sole surviving descendant species from the ranks of human-like hominids, which emerged abruptly in an evolutionary pulse after the australopithecines. The Pleistocene was an identifiable epoch of geological transformation in which the most recent forms of mammals evolved. It lasted until around eleven thousand years ago, ending with the most recent ice age. Its colossal ice sheets covered at one time all of Antarctica, large parts of Europe and America, and small areas of Asia.

There have been many glacial periods in the long history of the earth. The advance and retreat of the continental ice sheets has been interspersed with million-year spans of comparatively mild climate. The early humans, living as small, interdependent bands of hunter–gatherers and within farming populations, had to contend with drastic temperature fluctuations and natural disasters such as drought, flood and geological upheavals. Humans had spread out of Africa to parts of Asia by the early Pleistocene, about one and a half million years ago, and by fifty thousand years ago they had migrated to all parts of the world.

This dispersal diversified the human gene pool only slightly, even as the different populations began to develop extensive physical variations in skin colour and facial characteristics, which would later on lead to the development of distinct racial groups. Far less genetic diversity is found in human DNA compared with that found between chimpanzees and other closely related primates. All human beings have virtually identical DNA, with only small variations between populations. Humankind has remained a single, undivided species.

In new locales populations need to cope with conditions in various ways. Traits that confer an advantage come into their own and spread in the population. Genes can migrate in and out of a population, but geographic isolation is a buffer to gene flow; an isolated population is less likely to interbreed with other populations, so its gene pool is less open to inflow and outflow effects. Local ecological conditions may alter the adaptive merit of existing genetic variations. Where you live can affect how you look. Certain features – like colour of skin, eyes and hair – result from adaptation in geographic isolation.

Different races are ethnic populations of the same species, with no particular thresholds of biology, and all human beings belong to the same species, with features like skin and hair colour that emerge through variations in gene expression. Human geographical isolation did not split these races into subspecies or separate species. The genetic makeup of all human beings is thought to be more than ninety-nine percent identical; the difference between individuals even of different races is put on the order of a tenth of one percent. Human beings, therefore, are genetically very similar.

Today's human population of roughly seven billion, it is suggested, might have grown from a number of mostly sub-

Saharan ancestral groups. The gene pool lacked any large diversity then and it still does so today. We are, in our modern forms, genetically very close to our African ancestors. Chimpanzees, which share about ninety-nine percent of our functional DNA, are much more genetically diverse; they are a much older species, and have had more time to accumulate genetic changes and they have specialized a great deal since. Yet the life of a chimpanzee has hardly changed from that of its ancestor of some twenty thousand years ago; in contrast, patterns of behaviour among humans have been completely transformed over the same period.

As we have said, migration from the African homeland resulted in diverse human populations in emergent cultures. Certain accentuated variations in the common gene pool accounted for the evident racial characteristics, as the body adapted to new habitats and new diets. Some traits are clustered in definite regions. Asians, Africans and Europeans have distinctive skin colours, body shapes and facial features. Highly pigmented, dark skin is tolerant to strong sunlight, protecting against the sun's ultraviolet rays, strong exposure to which may cause skin cancers and impair production of folic acid. These ultraviolet rays also facilitate synthesis of vitamin D in the skin. In temperate zones, fair skin became capable of synthesizing vitamin D with adequate exposure to sunlight. The bodies of people in Europe took on a more compact form to conserve heat in the cool climate. In hot, arid conditions the shape of the nose and overall body shape changed to maximize the body's surface area and thus prevent overheating. The adaptive importance of some racial traits is unknown and may relate to differences in internal body processes.

Humans vary more within races than across races. For example, the battery of genes responsible for skin pigmentation exists in all the races but at different frequencies in different populations. There is variation in adaptation even among individuals in any population. Not all of them will reflect a trend identifiable in the majority. The same is true for blood type, hair and eye colour and immunity, and so on; they are all distributed throughout all the races. Traits can be acquired without any particular reference to race. Indeed, a person may share more traits with groups other than his own. In an age of high-speed transportation and global social connectivity via the internet, together with mass migrations and inter-breeding between ethnic groups, racial differences are a mosaic of linguistic, social and cultural elements, rather than physical and genetic traits.

In fact cultural evolution and historical change – more than genes – account for the diversity found among human beings. Customs, norms and habits, tools and language, social institutions and indigenous world-views are passed on from one generation to the next. Cultural development is rooted in an intuitive sense of connection with one's native soil.

Pale shadow

Diversity is the driving force of evolution. Charles Darwin (1809–1882) advanced the Theory of Evolution. He explains in his book *On the Origin of the Species,* which was published in 1859, that new species and their varieties emerge from antecedents through the struggle for existence, natural selection, and selective breeding. He surmised that humans had descended from African apes, or, more precisely, an ancient, extinct ape-like ancestor of both. According to Darwin, we are just a tiny part of a vast changing world. Everything is moving all the time. To live and survive animals and plants must adapt to changes in their environment, and they do so by small inheritable changes. Each habitat poses its own challenges, perhaps being very cold or wet, or hot or dry. To be able to find food and water, therefore, or shelter from the cold or heat and protect themselves from predators, all creatures need to develop survival strategies.

Genes change randomly, and adaptation depends on modification of a pre-existing form, from which future forms emerge. No longer is each creature occupying its space and place as assigned by providence; rather the driving force of nature's selection process, unbidden and unpredictable as it is, commands each creature. Darwin himself gained insights into the cause of evolution by observing domestic breeding of livestock and crop plants by trial and error manipulation, a process in which man is the agent of selective improvement of yields (at least, in their usefulness to him).

Darwin deduced that living things faced a constant struggle to survive. The population can vary in any number of ways, with some individuals thriving and leaving descendants, while some die without breeding. Some traits increase in frequency. This process is what Darwin called the principle of natural selection.

Excessively large populations are forced into competition for food, light, water, shelter and territory. A shortage of resources means that some individuals will die, that some populations will vanish and that some species will become extinct. We see this fierce struggle every day, both in the jungle and in the back garden (not to mention in the High Street). Neglected gardens become choked with weeds, and the same struggle goes on in the human jungle. Darwin declares: 'Man in his arrogance thinks himself a great work, worthy of the interposition of a deity. More humble and I believe true to consider him created from animals' *(Notebooks)*. For Darwin and his intellectual heirs, nothing is preordained about humankind coming into existence. Darwinian explanations crown our behaviour and mental capacities on learned associations acquired in the course of our evolutionary history.

Science is a wonderful servant to man, but it is a poor master. It tells us little about the conscious interactions of our mind or its relationship with the world around us other than elaborating on the information-processing and neurochemical computations of switches and relays, transistors and integrated circuits formed by the neural components of the brain. The brain processes our sensory perceptions, thoughts and emotions, and orchestrates our actions. Its patterns of electrochemical activity, however, reveal nothing about what it means to see and smell a rose, or hear a symphony. This is irrespective of whether our switches are turned on or not turned on, or whether these 'switches' are inactive in certain environments (physical or cultural).

While we must give credit to the Darwinian thesis of persistent change in the natural world, culminating with the large-brained mammal that is man, nobody subscribing to Darwinism nowadays would endorse all its assumptions as proven, or accept that Darwin worked out everything about how species develop.

Voyager of the mind

Darwin's ideas on the evolutionary process did not come out of nowhere. His insight drew on studies of fossils, geology and mineralogy. Evolutionary thinking pre-dated Darwin and the Christian era as far back as the times of the ancient Greeks, who conceived the idea of a purpose in the weaving of nature. Heredity and the links between various life forms were subjects in the minds of many biologists and philosophers in Europe from the eighteenth century onwards, together with a philosophy of gradual development. Naturalists had in a number of ways anticipated Darwinian concepts, even if they had not alighted on ideas about natural selection. Moreover, for centuries biblical representations had overshadowed attempts to make sense of the natural world. Some evolutionary models anticipated the cornerstones of Darwin's theory. Nevertheless, clear articulation of the ideas was still needed. Darwin postulated a plausible mechanism which, combined with the analysis of genetics, outstripped the interpretations of the other naturalists of his time, except for one Alfred R. Wallace, who had arrived at the same idea of natural selection quite independently.

The first full-blown model of evolution belonged to the French naturalist Jean-Baptiste Lamarck (1744-1829). His theory of transformation, published in 1809, attributed to organisms a 'will to development' together with 'an intrinsic tendency to strive towards perfection', whereby their needs shape their characteristics and behaviour, and changes in the environment cause changes in behaviour. Lamarck further advanced ideas about an ability to adapt to the environment through the inheritance of acquired characteristics. He even suggested that animals could transform into other animals; for example, he proposed that the giraffe developed its long neck by generations of ancestral animals stretching up to reach for foliage. This view was for a time also endorsed by Darwin. However, the idea that traits acquired in the course of an individual's lifetime can be passed on to its offspring was thoroughly disproved in the late nineteenth century with the recognition that germline cells, which give rise to the eggs and sperm that fertilize and develop to form offspring and propagate the species, develop quite separately from the other cells of the body (the somatic cells that make up the individual).

Darwin stressed the concept of natural *selection* rather than *transformation*, based on his observation that populations remain stable in size through the generations. Also, he noted that a single species could exist in a considerable variation of forms under different ecological conditions, and that a tremendous diversity of species inhabited identical conditions. This meant that these species could not have simply evolved by direct adaptation to the environment, as Lamarck suggested. Competition – in the form of a struggle for survival – means that individuals possessing particular variations thrive more than others. Such variations are likely to be preserved because that individual will live, and therefore its genes will spread in the species over subsequent generations. This process of natural selection became known as Darwinism.

Early Darwinism did not distinguish between genetic variations, which can be inherited, and *acquired* characteristics, which cannot be inherited. The discovery of genetic mutations overturned nineteenth century ideas of inheritance. The mechanism of adaptation came down to a tangled web of random genetic variations. The best-adapted survive, until they get weak or old and infirm, and are replaced by a fitter generation; the weak fall by the wayside.

Too solid flesh

Working out the mechanism of inheritance, of how variations arise, was left to the thorough work of one of Darwin's contemporaries, the Moravian botanist–friar Gregor Johann Mendel (1822–1884). He suggested that traits are inherited from one generation to the next through discrete units of heredity that were later called genes. The principles were drawn from his experiments with plant hybrids, leading to the emergence of genetic studies. In his mechanism, variations arose from genetic mutation.

After Darwin, people no longer saw themselves in the same way. Darwinism challenged the most cherished beliefs about the human self and the place of humans in nature; it struck at the heart of biology, religion and politics, and challenged the foundations of our behaviour and morality, as well as our interpretation of the world and history.

Darwin knew nothing of the as yet undiscovered molecule of DNA, and the chemical material of genetic information whereby the replication of DNA determines inheritance. The connection between inheritance and the gene was made early in the twentieth century, in the wake of Mendel's discoveries. Darwin's selection mechanism became intermeshed with the genetic relationships observed by Mendel. Adaptation through selection in evolution is supported by modern genetics.

While adaptation is an essential strategy for survival, the characteristics acquired in the course of life cannot be passed on by genetic means. The once-scrawny blacksmith cannot pass on his brawny arms to his offspring because the body cannot encode an *acquired* characteristic.

In the nineteenth century, the belief in improvement and progress through conflict and competition bordered on religious dogmatism. Darwin grappled with the serious pressures of the social ferment of exaltation and abasement in everyday living caused by the industrialized Victorian society he was living in. In this society, religion and science engaged with each other in a complicated alliance. He saw, in the struggle by everyone around him for existence in competition for limited resources, the dire implications of human overpopulation. The novels of Charles Dickens and the works of Thomas Carlyle castigated that time and place for its godlessness and lack of charity. But Darwin the scientist observed selection at work in the form of natural processes driven by struggle, without the certainty of a God or a struggle that transcended the immediate occasion. He framed the view that nature operates through competition by sorting out characteristics in organisms that confer reproductive advantage; the best-adapted survive and help the species to move on.

Cycle of the ages renewed

The biblical narrative of Genesis, especially its unreflected understanding outside the context of natural law, was, by the nineteenth century, being stamped on by advances of science in the understanding of the natural world. The astronomical and planetary observations made centuries earlier by Copernicus, Tycho Brahe, Johannes Kepler and Galileo transformed our view of the heavens. Charles Lyell's *Principles of Geology* (1830-33) showed that the earth, with its evolving landscapes, was a great deal older than the six thousand or so years attributed to the scriptures. The record in the rocks was quite at variance with interpretation of the biblical account of the earth's origins. Slow and accelerated events of natural forces permeated historic time. Fossils showed that living creatures were not static – but changing; animals were already extinct, including species that had gone, some before the scriptures were written. Nature, in its diversity, was anchored in unswerving physical laws, and any phenomenon interacting with the physical world was bound by the framework of these rules. For new 'scientific man' it was nature, together with natural law, that was the driving force of reality.

Many less radical souls remained convinced that in every geological period there were natural disasters that wiped out life, and new species modelled by God emerged, each in its appropriate form and filling an ecological niche. An analogy was drawn with the resurgence of nature in the spring. Each species, although capable of minor adaptation, was basically unchanging. Geological process, thus the large-scale environment, was anchored in transcendent purpose.

But goal-directed or God-directed imperatives lessened in significance in the wake of growing radical questions about the natural world. Ancient geological upheavals were shown to have been caused by natural forces, resulting in vast climatic changes. Any trend towards a stable uniformity of species and a static biodiversity was interrupted by catastrophes that left mass extinctions in their wake. Over vast tracts of time, the earth weathered drastic changes through droughts, gigantic floods, erosions, volcanic eruptions, earthquakes, ice ages, and geological shifts as mountains formed and continents drifted, to say nothing of cosmic encounters with comets, meteorites and asteroids. Such massive forces were shown to have shaped conditions on the planet throughout its history.

With such natural upheavals, life seemed tenuous and arbitrary. Fossil records showed that major upheavals had rocked the planet. Species perished and some never recovered, but new ones then emerged and occupied new niches and new ecosystems. The great cycle of birth and renewal continued unabated. By subtle adaptations among the existing forms, further new species came about. Eventually a conscious being, *man*, appeared – one who could ask questions and, because of his inventive behaviour, could even to some extent transform his environment.

Today we can see that we were carried away by this power to modify our environment, and we have interfered in natural processes without understanding, merely to pursue our own ends. Environmental problems, which have been the catalyst for the downfall of great and centuries-old civilizations and societies in the past, are in our current realities worse than ever. We are at the point where our inconsiderate plunder of natural resources, especially of the non-renewable ones, coupled with inadequate restorative practices, is a threat to the existence of countless other species. The fine balances and interdependencies required by the ecosphere to sustain life on earth have been upset by our need – or simply greed – to, as we put it, 'harness the forces within nature' through science and technology. Oil, coal and natural gas propelled our rise. Oil spills, nuclear meltdowns, chemical fall-out and climate warming threaten not only our future, but that of all life on the planet.

Method without intuition

Darwin's evidence of variation came from observing the close relationships between similar species in diverse geographical localities. Darwin, at the age of twenty-two, embarked on a five-year expedition of extensive scientific observations (1831–1836) aboard *HMS Beagle* – sailing via Tenerife, Brazil, the Galapagos Islands, Tahiti, Tasmania and other far-flung locations – with a belief in the immutability of species. With a Unitarian background, he had been a prospective ordinand who looked forward to gathering geological evidence possibly in support of the Old Testament world-view. But through the influence of his cousin, William Darwin Fox, and sponsored by his Cambridge botany professor, he was guided by more than religious ambition and remained loyal to impartial enquiry through his travels, desiring to chart the natural world.

He compared diverse flora and fauna, on rocks and in reefs. He studied the native people of Tierra del Fuego and examined the fossil finds of mammals. He was struck by the common ground shared between different species, but also by the way their forms, behaviour and ways of life were adapted to their environments. For example, the closely grouped but somewhat isolated Galapagos islands in the Pacific Ocean were replete with evolutionary trends among their species of finches, iguanas and tortoises. Closely allied life forms had variations based on their isolation from each other, and were different from similar species on the South American mainland. Darwin recognized from the fossil evidence that the plant and animal world of past periods deviated substantially from the contemporary scene; that the earth had changed over geological timescales, resulting in the diversity and myriad adaptations of living creatures. These discoveries drew him towards a vision of an evolutionary universe that challenged even his own religious beliefs.

On the *Beagle's* return, Darwin marshalled his research notes and specimens to prove the selectionist Theory of Evolution. The far-reaching diversity of flora and fauna could only be explained by evolutionary adaptation, which was shaped by natural forces. The argument that creatures were not designed to perfection in a habitat, but subject to pressure to adapt to that habitat, required no Creator – as long as certain conditions were met, life simply emerged spontaneously in purely contingent circumstances from universal materials. Once life was present, organisms developed new levels of form and organization of ever-increasing complexity.

Survival (of the species rather than the individual) was the result of what Darwin termed the universal struggle for life. Today we know that this struggle takes place not only at the level of species but also in the gene pool. No two organisms are exactly alike, no two zebras, no two human beings. Each is unique in some characteristic way. The differences are underscored by both slight and dramatic genetic changes, boosting or diminishing the frequency and advantage of some traits.

In a population, genetic changes occur all the time, with just one gene out of tens of thousands possibly taking hold. 'The preservation of favourable individual differences', wrote Darwin, 'and the destruction of those which are injurious, I have called Natural Selection, or Survival of the Fittest'. Cumulative change leads to diversification, and perhaps might allow metamorphosis of one species into another.

As it did for Darwin in his later years, the theory of selection in evolution satisfies us intellectually, but for us it is at a dulled remove from the complex individuality of human life. Science and religion, for thousands of years husband and wife, have in the past couple of centuries become divorced through deepening mutual suspicion and incomprehension. Religion has ceded her crown to science, and left the lead in science's hands – and he has gone on the rampage over her cultural preserve.

Manifestly, objective scientific scrutiny, the advance of scientific knowledge, has done wonders for humankind and made the world a different place. But there are greater truths that inseparably heighten our awareness and cannot be proved by the brain or recognized through the force of reason alone, but only in the heart, felt in the metaphorical pulses that reach us from immeasurable heights. We would be wise to acknowledge them. We have become obsessed with quantity and price, at the expense of quality, worth and value. To know something's price is useful, but to appreciate its value is to recognize a deeply human need.

Rhythms measureless and wild

In Darwin's time scientific thought accommodated God's benevolent actions in the natural world. God created Heaven and Earth with the seas, and the planets, the sun and the moon and stars. He created animals and the masterpiece – Man. God equipped species and organisms for the challenges of the environment.

The first half of the nineteenth century lacked a plausible theory of the organic change evident in the natural world. Natural selection, when it emerged as an explanation for the origin of the species, challenged belief in a Creator, but it could not explain the origin of life.

Darwin's research was logical and thorough. His innovative contribution to science demonstrated that the complexity of natural phenomena could be organized by the self-assemblying property of simple forces that did not require the personal intervention of God. Natural selection, he maintained, 'leads to the improvement of each creature in relation to its condition of life'. The contingent changes are not grounded in the needs of an organism. A trait, or structure, can be classed as an adaptation without any evolutionary advantage being inevitable.

Species change and transform themselves, and may become extinct, but there is no evidence for the transmutation of one species into another. The winged horse belongs in mythology. It is axiomatic that qualities must first exist before they are inherited. Amid patterns of repeats and variations, species can remain stable and little changed throughout their whole span of existence, and many survive relatively unchanged for millions of years, but new species can appear suddenly, in forms that are close to their final forms. The raven, one of the oldest species of birds, is still a raven. The cockroach dates back to a time before the dinosaur, according to fossil records, and it is still a cockroach. However, change can appear relatively rapidly in geological terms, with chemical-resistant insects, microbes and weeds and in fruit-flies and mammals, and still be gradual genetically. Evolution goes on at various speeds; we wrongly assume that it only happens over vast amounts of time. Sometimes a great change can manifest in a few generations, for instance when a small isolated population suddenly faces new ecological challenges in its environment. It is also possible to transfer animals and plants to an alien environment and observe, over a number of generations, some relatively rapid changes.

In fact, new species develop much faster than was once thought. They change, as if under time-lapse photography, whenever their environment changes. Rapid changes in characteristics have been noted, especially today, as climate changes occur far faster than ever before, together with accelerating loss of habitats and the potent rearrangement of relationships among life forms caused by human activities. Living things have an enormous and often underestimated potential for adaptation. Abrupt environmental changes can bring about a spurt of variations, not limited by a slow variation in the gene pool.

The fast and furious evolutionary changes that have been recorded in species of tropical lizards and fishes such as guppies would have amazed the Darwin who studied finch populations on the Galapagos Islands. The interplay of selection and genetic drift cannot fully explain the speed of some evolutionary changes.

As much as the Theory of Evolution explains many observations of species and is a pillar of science, specific questions of great importance remain unresolved if we do not simply submit to its doctrinal authority. It is an unproven assumption that evolution is the outcome of chance, or that no intelligence wills movement towards purposeful attainment of specific goals in the natural world. If that primitive life form found in Paleozoic fossils links a microbe to conscious man, who is to say that those massive evolutionary strides over eons are simply fortuitous and not subject to a higher will?

Selection is regarded as a means to adaptation. Darwin saw, for example, an advantage in the anomaly of the flightlessness of the ostrich, the largest living bird. Their grounded flock live in the company of grazing herds such as zebras, elephants and antelopes, often forming a mutual pact of alertness to danger from predators. Their stunted wings are the legacy of disuse, and render them flightless. Fossil records date their flying ancestor back some five million years, showing that a species can survive for thousands of years despite adaptations that go against the 'grain' of the species.

The flightless, waddling penguin has survived for fifty million years, adapting its undersized wings into flippers for agile underwater paddling and into weapons for defence. However, penguins have true affinity with ducks, which can fly.

Poised precariously between reason and emotion

Darwin's speculation on the descent of *man* has little to say about how an ape-like ancestor can develop into a moral being with a sense of selfhood, with attributes such as reasoning and intelligence, language, ingenuity, inventiveness and creativity, and folly and ambition. Darwin concludes that the categories of our morality have developed more by habit and training than by natural selection. According to him, we are just a further stage of selection from the great apes; our attributes, language and consciousness for example, are merely on a scale that is greatly magnified from the amoeba, by way of the insect, fish and duck to the great apes. Man is to be seen on the same continuum as the beast. Contrary to Hamlet's exaltation of man as a piece of work that is 'noble in reason! ... infinite in faculty! ... in apprehension like a god!', our conscious mind, Darwin concluded, was developed from 'a mind as low as that possessed by the lowest animal'. 'Man with all his noble qualities ... with his godlike intellect which has penetrated into the movements and constitution of the solar system ... still bears in his bodily frame the indelible stamp of his lowly origin' *(The Descent of Man, 1871)*.

Even today such an unashamed dethronement of man to brute is mildly disturbing. For Darwin's Victorian audience it would have been an earth-shattering blow. It is then little wonder that for two decades Darwin kept his thesis of natural selection to himself. The shockwaves from the impact of his theory are still being felt today. Once godlike, now brutish, Victorian man was benumbed by the rather extreme alienation, and no longer knew where he belonged in the scheme of things. Faith slipped its moorings in a retreat of religion, together with the concept of revelation. Thinkers such as Darwin, Thomas Henry Huxley and Herbert Spencer were even beginning to suspect that there was no cosmological blueprint. Tennyson's long confessional poem *In Memoriam* of 1850 begins very positively: 'Strong Son of God, immortal Love', but it is interlaced with doubts about the Creator and the nature and origin of man – so much so that the poet was forced into the solecism of confessing: 'There lives more faith in honest doubt, Believe me, than in half the creeds'.

By the middle of the twentieth century, full-blown doubt not only had the upper hand, but it also contained the metaphysical traditions and would frame the cultural mores in all-out atheism. We had gone from *believing* everything to *doubting* everything. The neo-Freudian Erich Fromm pronounced in *The Sane Society* (1955) that God and man were dead. Yet thinking man's emotional needs and his spiritual yearning to transcend the universe remained. We wrestle with the fundamental mystery of our condition, in which the tension between doubt and truth is expressed with profane ludicrousness by the blind and crippled tyrant Hamm in Samuel Beckett's play *Endgame* (1957). This was a nihilistic portrayal of the dimmed insight of the human mind together with an equivocal hope of redemption. As a non-believer, Hamm was at the end of his days and laughably destitute of hope, but he fell back on offering up a prayer to the Almighty, only lugubriously to deny His existence.

Today we understand the world around us very differently from our forebears. The Copernican revolution shifted us from a privileged central position to the fringes of the universe. Nature is indifferent and could easily discharge us. We now know that the earth is but a tiny particle of dust in a cosmic expanse, made up of the same materials as the stars, and that the atoms in our bodies are recycled from generations of stars. Each galaxy is strewn across an expanse of thousands of light-years and comprises hundreds of billions of stars (one light-year is the distance light travels in a year). And the visible universe contains more than a hundred billion galaxies. Little man is merely a speck of dust in the cosmos with the idea of a Great Creator existing as a shadowy concept in his mind.

While Cambridge University was closed during the time of the Great Plague in 1664 and 1665, Isaac Newton studied the nature of light. His experiments involved passing light through various substances and glass prisms, splitting white light into rainbow colours, and showing that rays of light differ in their colour and their ability to refract or bend. He concluded that the indistinct rays of light incident upon the object-glass of the telescope were being brought to focus at different distances, leading to his theory of 'differently refrangible rays'. This led to the development of a far more accurate reflecting telescope by William Herschel. Newton's later descriptions of the invisible force of gravity pulling heavenly bodies towards each other, brought the heavens down to earth and consolidated our outlook on the world as a machine, with God as the Almighty Clockmaker.

But it was Darwin's presiding ideas about evolution that put God to rest in the imagination of many cultures. Now man was not only just an animated speck of stardust dangling in the cosmos; he was also a hairless ape who walked dully along, with no unique cultural attributes in the struggle of living matter to survive. Like the animal, his destiny has to run its course, controlled by chance. Chance rules the world. Improvement comes through struggle. Diverse species exist, adapted to all possible ecological niches. Some 'lucky' species survive; others fall by the wayside. If everything is left to the drift of time, the right living things emerge, and what lives, lives. We are just as likely, in the Darwinian scheme of things, to disappear from the earth, like more than ninety-nine percent of all species that have ever lived. That is, if we do not eliminate ourselves first.

Darwin never denied the existence of God and he came close to believing in the immortality of the soul. As he said in his autobiography, 'Believing as I do that man in the distant future will be a more perfect creature than he is now, it is an intolerable thought that he and all other sentient beings are doomed to complete annihilation after such long-continued slow progress. To those who freely admit the immortality of the human soul, the destruction of our world will not appear so dreadful'.

Science teaches us one supremely valuable lesson: that what we see is not all there is. That looks can be deceptive. For instance, the gravitational pull that holds the hugely diverse galaxies together points to the transcendence of gravity. The formation of the physical world around us is governed by what we do *not* see.

We appear to have moved from one creed to another. Having escaped from orthodox religion, which the nineteenth-century novelist and 'freethinker' George Eliot (Mary Ann Evans) branded as 'this wretched giant's bed of dogma', we simply embraced faith in the new orthodoxy of scientific secularism for truth and discovery to interpret the world, and in the process we dismissed the otherworldly dimension of consciousness as a superstition of the lazy mind or as an attempt to bamboozle thinking man. Science is the faith of the age of reason. We are led to believe that ours is a mind ruled by reason as the fundamental attribute of our ability to understand, and by the 'self-interest' of our genes. We share a genetic code of just a few chemical letters that is universal and can be copied together with the self-assembly property of DNA from site to site, in the developing embryos of utterly different-looking species that diverged evolutionarily a very long time ago, and ultimately thereby life is introduced into almost every available niche on earth.

But surely we know better! We feel that conscious mind, selfhood, moral reflection, culture and civilization are purposeful to our being, not sidelines to a creature adapting to its environment. Science deals with what we *know*; art and religion with what we *feel*. Is there a filtering interface of sorts between them? We cannot account for some things in our thoughts, but we value them and they define us as human.

As for the creation of the cosmos, should we delete God and substitute the primordial 'big bang' fireball instead? If the ancient conception of the gods or God was illogical, the substitution of the fiery colossus of an arbitrary 'big bang' creation without a preceding energetic will is even more so. We, as part of nature, abhor a vacuum as nature does. Today, in place of God, there is the Big Bang. Tomorrow there will be something else!

Science, as we have said, is a marvellous tool, and one that is being constantly honed and refined in its theories and belief systems. Although a good servant, it is a bad master. Our science, of which we are rightly proud, will never verify reality to its highest truth if every 'discovery' continues to be circumscribed to material processes, dug from below and constrained by the time and space conceptions of the intellect. The intellect is only one pole of our consciousness and is subordinate to our intuitive recognition.

TWO

Gene

The argument about our origins is wedged between the idea that we are pre-ordained with a God-given purpose and the thought that we are descended from animals, in particular from the great apes, and that our nature arises in large part from our DNA blueprint. At the heart of the argument is our true core of reality. Philosophers and biologists, in trying to fathom what specifically being human is, at first looked to the genes for the salient information, then to the brain. Genes, invested with almost magical powers, lurk as our tiny puppet masters, with the prospect of transforming us into characters of moral equivocation such as Lilliputians, Houyhnhnms, Yahoos or Chirons. Genes tell us nothing about free will or the emergence of our conscious self-awareness. At stake is the authenticity of our human freedom and self-fulfillment.

The gene is the functional unit of inheritance. Genes flesh out the body plan, from humans to algae to blue whales. We and the great apes share a striking similarity in our genetic makeup, the legacy of a common ancestor that lived millions of years ago. Our early hominid forebears went one way and adapted to life outside the forest; ancestors of the great apes carried on in the lush environment of the forest.

The multi-billion dollar project to map the one hundred thousand or so chromosome sequences of our genetic make-up has partly been motivated by great curiosity about what it is, in relation to the genes, that sets us apart from animals. As the view that we have some kind of separate soul fades in the western world, and the notion of God recedes in the consciousness, modern science steps into the vacuum thus created, on the one hand to verify faith, or, if not, to sort us all out as mere material at the core, putting our cognitive modernity down to some feisty little beads on a string.

Beads on a string

Genetic material is packaged in a storehouse of information in the nucleus of each of the approximately one hundred trillion cells of the adult human body. Chromosomes are thread-like structures in the nucleus, containing a long strand of tightly coiled DNA, which in turn contains thousands of encoded genes – analogous to the beads on a string. Each gene comprises a sequence of hereditary components of the DNA, which is wound tightly within the chromosome to accommodate its considerable length inside the cell's nucleus. The genetic information is a sequence of interconnected chemical (nucleotide) bases in twinned spiral strands that twist around each other to form the double helix of the DNA molecule. It is supported by a protein scaffold of structural proteins that also controls gene expression. The chromosome is the physical carrier of heredity, and the gene is the information unit. Genetic information encoded in the structure of DNA, when set adrift, operates like a blueprint for our physical appearance and explains why we look like our parents.

Human chromosomes range in length and occur in homologous (matching) pairs for a particular characteristic or trait, half derived from the mother and half from the father. Each chromosome contains between hundreds and thousands of genes, and their nucleotide components encode the thousands of proteins in all parts of the body. Proteins are molecules – a molecule is the smallest unit of a chemical compound that can exist alone (an atom is the smallest unit of an element that may or may not exist alone). Proteins are the workhorses of the cell, with an all-purpose role in the body: they are active in the production of most biomolecules, including other proteins, they regulate cell function, they build the body parts and provide structural support, they shape the body's physical characteristics, make muscles work, transport oxygen, build the components of the immune system and are constituents of enzymes and hormones.

A protein molecule is a large complex, made up of various amino acid subunits – of which there are about twenty. A trio of chemical nucleotide bases on the DNA strand forms the code for specific amino acids. An average gene contains three thousand bases – the blueprint for a protein of a thousand amino acids. In any one species, two individuals will differ by only twenty to thirty bases.

Genes are passive. In themselves, they can do nothing. They have to be activated. Genes are about their products, and the business of linking cells and organs through feedback loops of activation and inhibition. For a gene to be expressed, however, its DNA segment in the chromosome must be unraveled and exposed to a complex of protein enzymes. A missing gene results in an ineffective product or none at all. A single gene mutation – a single base difference in the DNA – causes inherited anaemias like sickle cell disease and thalassaemia. Diabetes is another disorder with a genetic basis. The hormone insulin is produced after a meal to regulate the level of glucose in the blood. It affects how much glucose is available to flow into the cells from the bloodstream, as well as its storage as glycogen or fat. In diabetes, the output of insulin is abnormal. The defect is from the interaction of multiple genes and the environment on insulin production, glucose transport, and insulin binding to receptors on cells. More than a thousand human diseases are attributable to an alteration in a single gene.

The six-foot length of DNA in a human cell nucleus is only billionths of an inch wide, and consists of about three billion nucleotide base pairs. The entire DNA in all the cells of a human body would stretch out if placed in single file from the earth to the moon and back a hundred times, with only a tiny proportion of that length containing protein-coding genes. Discrete DNA sequences of protein-coding genes – less than five percent of the human genome – are clustered like islands in an ocean of non-coding regions. The roles of the large proportion of non-coding DNA include providing structural integrity for chromosomes, regulating protein synthesis, and possibly compensating for mutations. Up to twenty percent of the DNA has no known function.

Count every dandelion

Physical characteristics are passed on from parents to offspring by means of genes. Human heredity resembles that of other life forms. The entire human DNA complement of genes (the human genome) has far fewer genes than was assumed and hardly any that are specifically human. Many plants, like the lily and the bean, have more genes than humans. As does the salamander. We appear to have about thirty thousand protein-coding genes, about as many as a mouse, which is little more than double the number in a tiny fruit-fly, and barely fifty percent more than the roundworm or the weed known as thale cress.

We share nearly ninety-nine percent of our functional DNA with chimpanzees, ninety percent with cows, and about seventy percent with mice. Half of the genes in a banana are also present in the human body, and we share the same genetic information carried by DNA and the mechanism of DNA expression with other creatures. The universal genetic code means a gene from one creature can be transferred and expressed in a different creature. Some human diseases can be reproduced in the mouse, and some even in the fruit-fly. Mouse genes can replace genes in the fly, where they will express fly and *not* mouse characteristics.

The genetic complement acts as an integral unit. Most genes are multi-faceted. It is far too simplistic to be talking about one gene for this trait and another gene for that trait. Factors such as conscious will and immediate circumstance influence gene expression as well, and in many instances add another layer of complexity to the correlation between gene and disease. Ailments result from an interplay between the environment and a number of genes. Moreover, investigators are finding out that, beyond genetics, life experiences can chemically modify DNA in chromosomes without altering the genetic code or tinkering with the basic essence of the gene, but still control expression of the information content. For example, there are many investigations that show a strong interaction between the body's immune defence system and the psyche.

Genes may accentuate or lessen each other's effects, for instance increasing or decreasing skin pigmentation. Blue eyes are not due to the presence of a blue pigment but arise from the lack of a gene for brown pigment. Each gene accounts for only a small effect. Genes also bring about unknown effects, which can make a huge difference to an expected outcome based on a few known genes. Gene families, clusters that evolved from one or a few ancestral genes, also have multiple copies and can sometimes tolerate the loss of one another's function.

The full complement of genes in the human genome is organized into twenty-three pairs of chromosomes. One chromosome of each pair is inherited from the mother and the other from the father. Each sperm, like each egg, contains only twenty-three chromosomes, half of the full complement. After fertilization the two fuse together and the zygote (derived from the Greek word for 'yoked') has a full set of genes, half from the father and half from the mother. One pair of chromosomes, the sex chromosomes, are matched in the female (XX) but mismatched in the male (XY). The other twenty-two pairs, called autosomes, are similar in males and females. Each cell in an individual contains the same genes, but different specialized cells express different subsets of the genes. Thus, a heart cell has the same set of genes as a brain cell, a skin cell or a hair cell, and so on, but selective gene expression differentiates one type of cell from another. Subsets of genes are switched on and off in the embryo to give rise to different types of cells, and over the course of life. A person could take on different physical characteristics from a variation in gene expression in his cells.

Genes exist in functional variants called alleles. Alleles have subtle base substitutions in their DNA sequence. Most alleles of a gene are harmless. One allele may produce black or blonde hair; one a brown eye colour and another blue. We normally possess no more than two alleles of a gene. Genes are dominant or recessive. Alleles pair up as co-dominant, co-recessive, or as a dominant and recessive pair, with one allele from each parent. In the ABO blood group system, for instance, A, B and AB are dominant alleles and O is recessive. Every person carries a number of severe genetic defects from alleles inherited from one parent, which are overridden by a normal allele inherited from the other parent.

Fostered alike by inheritance

No two people look exactly alike – not even identical twins, who have virtually the same DNA. Children will carry genes that vary slightly from those of their parents, and sometimes a parent and child have little obvious resemblance to each other. In a gathering, some people will be tall, some slim, others stout, some will have brown eyes, others blue. Yet we are all very similar in our genetic makeup. We look different from one another because of slight variations in our complement of genes, and because our life experiences can alter the expression of our genes. The tiny differences in our genetic makeup reside in two to ten million out of the three billion DNA base pairs, much less than one percent as genes go. The differences are due to genetic recombination during sexual reproduction.

In genetic recombination, DNA fragments are exchanged between chromosomes in the formation of sex cells, namely the egg (ovum) and sperm. Crossing-over of broken-off, matching segments between parental chromosomes generates new genetic combinations. In addition, each chromosome is independently assorted into the sex cells. This means that more than eight million chromosome combinations are possible in the egg or sperm and up to half of the coding sequences are unique in the individual genome. Certain genes tend to stay together during recombination. Such linked genes are co-inherited. The closer together the genes are on the chromosome the more frequently they will be transmitted together. Genes that lie far apart are more likely to be segregated onto different chromosomes.

The fertilized egg divides several times into an embryo. Stem cells are immature cells produced in the first few divisions, which have the capacity to divide into multiple cell types and the potential to repair most tissues in the body. Some become nerve cells, some develop into muscle cells, and others give rise to skin cells and blood cells and so on. Cell division involves turnover of certain molecules, mainly proteins and nucleic acids. Most cells divide throughout life. The body requires daily the provision of billions of copies of some fifty thousand different proteins, and about half the body's proteins are replaced every fortnight. Protein synthesis is highest in the brain, with twenty percent of an individual's energy and more than half of the genes being devoted to the cognitive landscape of a brain – an organ that is only two percent of the weight of the average body and weighs in at just a puny one and half kilograms (or three pounds).

Private presence

We share about ninety-nine percent of our same DNA with chimpanzees, as well as our ability to use tools, and we have remarkably similar fingerprint patterns. Chimpanzees can manipulate symbols in a language-like manner, showing rudimentary abstract thought processes and an ability to string two or three words together like a two-year old child. Some are also capable of being cunning and deceitful. Of course, the intellect is not unique to us. Most animals deploy skills in using tools, cracking nuts or finding and storing vital necessities. Many show tremendous cunning in their survival strategies. These attributes, however, have no affinity with our distinct capability of self-awareness.

All conscious organisms process information, and they do this with some capacity of perception or thought. This can range from simple, fixed-order movements and reflexes toward stimuli, purposeful instinctive action and the capacity for conceptual thought, right up to the capacity for language and self-knowledge. A simple step of consciousness is being awake and reacting suitably to stimuli. Thus the amoeba turns away from a source of irritation or discomfort.

Instinct is a natural and unreasoned pattern of behaviour, a predisposed impulsive action. In native fear, a rat will 'freeze' in its hole at the scent of a cat prowling outside. It may remain immobile, assessing the active threat and hoping the cat will wander away, or it may slink away if it has an escape route.

Animals achieve cognitive feats, some of which are comparable to those of a one- to two-year-old child. Parrots, chimpanzees and dolphins, for example, have a capacity for thought and can grasp some higher concepts and complex relations. Some animals plan with a degree of flexibility and exhibit rudimentary powers of deduction. Every pet or animal lover knows that animals feel and express joy and pain. The love of an animal towards a good master can be unbounded and unconditional. But although a creature is subject to the same natural laws as us in the interconnectedness of nature, the animal is not *self*-aware; it might be conscious of its own existence, but it is not aware of itself as an individual. Shame! *Tales of the Riverbank* always made many a person believe the contrary!

Our sense of selfhood has evolved far beyond the robust kinship inclinations of animals, and with a spur to culture in which contractual settlement underpins social ties. While we perceive the external world through our physical senses, and our mental landscape registers our insight in the form of feelings, we have a higher faculty of intuitive perception in the form of a deep inner automatic recognition, an immediate certainty without rational analysis. First come the deep inner feelings, then the rationalization follows. It surpasses intellect and instinct. We have an unbounded urge to inquire and explore, to go beyond the satisfaction of physical appetites and yearn for unknown realms. We have ambition. It is almost impossible to imagine a sheep thinking 'I must eat all this grass before sunset'.

Creatures with highly organized, sophisticated brains are a late development in the evolving phenomenon of life. Primates and social animals communicate with each other by exchanging detailed symbols. Dogs communicate, as do ants, bees, dolphins, elephants, whales, apes and a great many other creatures, through a set of prearranged signals. 'We, too, make noises when we laugh or weep.' But we use language to convey perceptions and concepts. 'Words are for those with promises to keep' as the poet W.H. Auden observed. We grow mentally with our capacity to understand and use language.

For millions of years the brains of fish and octopus have operated without being capable of self-awareness in their patterned behaviour. The ancient subcortical pathways of the autonomic nervous system operate subconsciously to control automatic bodily functions such as muscle and tendon contractions during reflex movements, lung and chest movements during breathing, the heart beat, the constriction and dilation of blood vessels to maintain blood flow and pressure, as well as body temperature, digestion and metabolism. The subcortical brainstem structures join the brain to the spinal cord and relay signals from the brain to the muscles and from bodily sense organs to the brain. Cortical centres needed for imagination, thought and planning, as in humans, are grafted upon the low-level primitive structures, and the cerebral formation is restricted mainly to birds and mammals. The same brain areas are activated in humans and animals when experiencing feelings of joy and pain.

Six million years ago the most developed brain on earth belonged to the chimpanzee. Later came the hominid brain, with a complexity associated with more purposeful behaviour. Learning refines the circuitry of the brain, and the more voluminous and sophisticated the brain, the more imagination and foresight tend to be developed. Beyond the chimpanzee, parrot and other animals with a capacity for primitive thought and making simple observations, we have the facility of selective conscious attention, jumping from one train of thought to another, and shifting one perception to the next. Above all, we have also moved beyond instinct to intuition and moral virtue.

Nothing new in principle was added in the evolution of the human brain. The brain of birds, for example, is considerably smaller than that of the chimpanzees, yet the cognitive feats of the two species are comparable. A big brain is more likely to facilitate learning and acquisition of specialist knowledge than to grant overall intelligence itself. Albert Einstein, for example, had a smaller than average brain for a human.

The brain is essential for our daily existence. It is the physical interface of the mind and the mediator of our conscious awareness. It generates the thoughts, emotions and perceptions that guide our behaviour. Everything we think and everything we do feeds back to the brain, and in the process motivates or demotivates us. Every perception releases chemical compounds in the brain. Every thought activates nerve impulses that travel to the body. Our brain executes all our actions, whether in our learning, our aspirations, our drive to improve ourselves, in our desire for something or someone, in liking, amusing or trusting others, or in destroying or in being transformed into brutes. We have become very proud of, even obsessed with, our large brain and the things we can do with it. We admire the 'brainy' and delight in having 'brainwaves'. In some ways we have come a long way from the caveman in our understanding of the world around us. We have pushed back a little the frontier of space and mapped the genome, among other feats. In other ways, we are as primitive as our thousand-year-old ancestors. Moreover, the brains of elephants, dolphins and whales are as big or bigger than ours, but they have less cognitive ability.

The child is born with a large brain and skull. Painful childbirth is a consequence of the runaway increase in brain size, to which the woman's pelvic formation together with the birth canal, in terms of the relationship between the waist and hip, has not adjusted. A fetus in the womb sprouts neurons at the astronomical rate of two hundred and fifty thousand per minute, on average, outside the first few weeks of gestation when no brain structure is present. At birth, the newborn's brain has more than one hundred billion neurons. At the age of three, about one thousand trillion connections between those neurons have formed in the child's brain. This is many times the number of stars in the galaxy and twice the number of connections in an adult brain.

Missing part of life melody

The fetus, the unborn child, is, for legal considerations at least, accepted as being alive after it makes its first movements (called the 'quickening'.) This commences around the middle of pregnancy. How and when the embryo, the minute ball of cells that follows from conception, becomes an 'individual' is a contentious issue, and one that divides religion and science. The religious viewpoint aligns with the assumption that a soul is acquired at conception and all emergent aspects of the individuality of personhood are contained as a minuscule replica within the fertilized egg, which simply needs to expand with growth. The embryo is already considered to be a human being with rights enshrined in the nation's laws. The rival scientific contention categorizes the embryo as a dismembered germ, which requires further stages of transformation as encoded by the genetic blueprint. The complement of genes holds several possibilities for realization of an individual. The embryo is allocated to a non-human category, that is to say, a mere clutch of cells without personhood until the formation of brain structures. This process starts at about three weeks and heralds the onset of individuality. In other words, human individuality is formed in a continuing process rather than being preformed or the result of a particular event in development. This has opened the way for research into human embryonic stem cells and the prospect of utilizing embryos for therapeutic and cloning applications.

Cloning involves resetting the cell's imprinted pattern of gene expression to the state of the embryo. In therapeutic cloning, stem cells may facilitate treatment for a range of ailments such as diabetes, cardiovascular disease and neurological disease, and they may be used to correct hereditary diseases and debilitations resulting from accidents. Stem cells have the capacity to differentiate into any specialized tissue, such as nerve, heart, muscle, or blood. Therapeutic cloning may therefore help to relieve grave afflictions resulting from strokes and spinal cord injuries, to reverse neurodegenerative disorders such as Parkinson's and Alzheimer's diseases, and may even improve cognitive impairments. Disabled stroke victims might recover impaired or lost functions like walking and talking. Stem cells derived from an individual's own body (as opposed to an embryo) would carry no risk of rejection, and may even be modified by gene therapy to repair one's own DNA, so allowing the individual to be cured by his or her own tissue.

The process of cloning involves removing DNA from a cell and injecting it into an egg that has been emptied of its own DNA signature. It is possible to simply take an egg, remove its nucleus (a process known as enucleation), and inject the genetic material from a skin cell, for example. The egg is then shocked with an electric charge to undergo division and thus generate an embryo. The vexing issue of cloning stirs much alarmist speculation. Cloning is basically the same process as when we take a cutting from a shrub and cultivate it to grow into a new plant. It is reproduction without sexual fertilization. That is really all it is.

Certain technological breakthroughs and applications of science may seem unnatural, like some sort of tampering with nature, an example of man's hubris, or an attempt to play God. However, we regularly use nature for our own benefit, intervening in natural processes. Indeed, we could hardly survive without some small acts of 'creative intervention'. We cultivate gardens, we breed and domesticate animals and we educate children. Scientific and technological advances yield the latest drug and surgical treatments. True, in these cases we are not creating as much as pressing natural forces to our advantage. It is but one step to use nature to create a device or engine or structure, and yet quite another to create not a jet engine or a hydro-electric dam or an atomic bomb but a 'perfect, disease-free human being'.

We may pause and reflect on just how far our technical knowledge has progressed over the past century, with major strides and often unexpected results. We have gone from the internal combustion engine to the nuclear war-headed rocket, from discovery of the gene to germ warfare. We are certainly, as ever, 'tampering with nature', but with unpredictable results. Of course the very phrase 'tampering with nature' is a solecism, for nature embraces every atom, gene and galaxy. We can have no predictive control over nature, which is subject to laws – and our understanding of these laws, despite our amazing progress in some fields, is still in its infancy. We are, in the expressive power of our self-awareness, really just babes in arms seeking – like Ulysses – for we know not what. Only a human will, empowered by knowledge of the working of the cosmic laws, would grant predictability to the discoveries from our experiments. We would then, of course, also have the insight not to venture into fields that are best left alone.

Proponents of cloning point to the myriad likely benefits: infertile couples could have children, hereditary diseases could be stamped out, degenerative diseases could be treated and suffering alleviated, while cloned 'spare body parts' might be available to extend and enhance life, regenerative medicine might eradicate disabilities that we are better off without, and threatened wildlife might be preserved. Of course, each invention or scientific 'breakthrough' brings its problems. Sometimes nature brings about its own cure, and sometimes science proves harmful to human happiness – one needs only to think of the drug thalidomide.

Other ways than cloning exist to prevent or eliminate terrible hereditary defects. One is genetic counselling, together with predictive DNA testing. It helps to determine risk of certain disorders before conception. Another way is in vitro (test-tube) fertilization with genetic determination and selection before implantation to produce disorder-free embryos.

There are practical and ethical arguments against tampering with our genes, which constitute a key zone of our existence and where unpredictable genetic errors could affect what looks like a normally developing cloned embryo. We have no way of retracing the history of a particular gene. Nature will usually abort an abnormal fetus, but we cannot always determine what is abnormal until it is too late. Some genes are active early, others only become significant later, when the abnormality shows. To opponents of cloning, the many unknowns about stem cells and cloning outweigh the likely benefits of the technology. The gene pool may throw up physical and moral freaks. We may take DNA from the brain of an Einstein or a Shakespeare, but we have no way of knowing or controlling the outcome.

What is not really examined in the two-way split on this issue are the spiritual implications of human cloning. Cloning is limited to the body. There is no cloning of the spirit. It is true that we exist as a physical organism in a predictable way. Our essence is, however, not our predictable body, but our unpredictable spirit, our core of something other that makes us a free individual with the ability to choose, to some degree at least, how to live and for what purpose. The inner development of the individual is a process of ennoblement and self-realization through knowledge and experience.

Cradles to preserve as sacred

The living core of spirit animates our body, which has no life without the spirit. It is an implement for the spirit. The natural world changes by evolution and dissolution. Death and decomposition are followed by fresh union of material components. The human spirit is also subject to development, but without dissolution, which is characteristic of the material. The body dies and our spirit leaves the body.

Modern reproductive technology, including in vitro fertilization, is subconsciously following natural principle. In other words, it is simply adjusting the existing laws to obtain a good general benefit, stretching the rules, so to speak, for our own advantage. All the same, the birth of a child will require the act of procreation and the incarnation of a human spirit.

The body takes shape. The incarnated ego, the spirit, glows through in the person's character. The spirit core is enveloped by finer coverings that are nested in each other and connected through finer radiation energy. These coverings correspond in consistency to the planes through which the spirit, a germ in its original state, passed down from its transcendental origin to the earthly environment. Without the physical body the so-called 'soul' (the spirit with otherworldly coverings) is the form in which we exist before and after our earthly life. By analogy, we put on different clothing, according to the conditions around us: a topcoat for the cold in winter, a diver's suit for undersea diving, or a spacesuit for venturing into the frigid void of space. We appear different, therefore, in different conditions. The difference between a body and soul shows, for example, during dreaming, when the body is at rest. In a spate of incisive dreaming that lasts for only a very short time, the soul is able to loosen its connections with the body and wander in a different reality, where the concept of time is broadened. A minute of dreaming translates into a quick succession of intense perceptions, with some unconscious ethereal influences. Then, as the body awakens, the connection between the body and soul becomes firm again.

A human clone, like identical twins, which are nature's clones, will be a separate personality from the individual the first cell or cells were taken from. With cloning, the physical body is forcibly given a start, on the assumption that it will take on an inner life. Cloning may run to some plan with animals, whose animistic soul core is a different species from the human spirit, belonging as it does to a group soul. The human body, however, is merely a cloak draped around an individual core of spirit. Not that our spiritual aspect is subsumed into the subconscious, as is widely accepted, appearing to take over the conscious brain spontaneously in moments of unbidden insight in the course of life. Simply put, you cannot clone the spirit. Therefore human soul-cloning appears to be a cul-de-sac, a dead-end, in that clones, even if viable, will receive spirit entities completely separate from the cell donors.

Even animal cloning has been shown to result in serious abnormalities. Animal experiments produce late abortions, stillbirths or severe deformities that affect most of the clones. Nearly every cloned animal has some genetic or physical defect, and normal-looking clones may carry serious distortions in their gene regulation. Researchers have as yet no insight into why almost all clones are abnormal.

Public health is more likely to be enhanced by simple non-invasive DNA screening. This will allow preventive treatment to be tailored to the individual, with efficient use of health resources. There may come a time when certain hereditary abnormalities in the embryo can be repaired with healthy DNA, which may have been synthesized by genetic engineering or transplanted from healthy cells. We may even be able to treat diseases preemptively before any symptoms arise.

We now have the eugenic tools to change human and other creatures at will. We might come to redesign many traits among people, selecting the best from the genetic pick-and-mix. It is possible that within a generation or two, or a millennium or two, we might engineer a new type of human in the image of our own choosing. But is this the bright prospect of a new biological Big Bang, or is it the nightmarish crunch of a Big Chill?

THREE

Atom

We can only speculate about how many other planets among the countless number existing in the universe are capable of supporting life. We simply do not know whether the earth is a rare exception, or even unique, in this respect. Our planet is situated at just the right distance from the sun to receive adequate warmth and radiation to allow life as we know it to evolve and to thrive. A DNA process that dates back over three billion years is one explanation offered for life on earth. The DNA molecule has the capacity to replicate itself, but the copies are not always identical. They are subject to slight changes. Thus from the simplest of beginnings the great biodiversity in the natural world has come about, culminating in (at least it seems so to us) in the large-brained hominids – human beings!

Creatures change and evolve over many generations. Fossils of primitive microorganisms date back some three and a half billion years. DNA has been recycled through the twists and turns of billions of years of evolution to produce plants, dinosaurs, whales, apes and humans. This being so, our life, and by implication our death, assumes the eclectic dispensation of genes that can morph from lowly bugs into people!

Genes are responsible for the reaction of an organism in an environment. Genes encode the blueprints for producing proteins, and proteins control chemical reactions. Some researchers conclude that early in its life, our planet offered a primordial soup of non-living matter, which transposed into self-replicating chemicals. Others speculate that the building blocks of life could have been seeded by elements delivered from deep space on cosmic dust and meteorites.

Life seems to us not only a most chancy phenomenon but also an extremely fragile one. More than ninety-nine percent of the species that have ever lived on this planet have become extinct. Dinosaurs and most species were wiped out at the end of the Cretaceous period some sixty-five million years ago, as the continents took on their modern forms. After enduring for one hundred and fifty million years, they became extinct. Quite possibly there were massive climate changes caused by dust and gas dispersal from huge volcanic eruptions or the collision of asteroids or comets with the earth. Some global cataclysm may have spread fire over the earth, so that smog blocked the sun and prevented photosynthesis by plants. Or an ice-age cycle might have developed from causes yet unknown. The end of this dominance by reptiles on land and sea opened the way for the mammals to predominate on earth, thereby ushering in a new era in the evolutionary cycle, one that culminated in the appearance of our early primate ancestors.

Bright shoots of regularity

Since the dawn of civilization, man has been animated by questions about the origins and structure of the cosmos, how it came into being, and what it primarily consisted of, when time began, and what future awaits it. Our world picture changes dramatically every few years, remodeled for each era, becoming at times more of a kaleidoscope than a static picture. Sometimes, however, the picture remained pretty stable for a few decades. This is the case with the current Big Bang theory, that cosmic expansion in time from a dimensionless point of infinite density that resulted in extremely hot, dense, high-energy initial conditions. The temperature cooled as the universe expanded. Galaxies, suns and planets formed, and (on at least one planet) life as we understand it began.

What happened before the Big Bang is in the realms of the unknown, beyond our barrier of thought. We cannot know about thoughts that we cannot think; we see only what we know, and we do not know what hinders our insight. It is as if science acknowledges the bounds of the intellect here. Cosmologists think that other 'big bangs' might have preceded our own Big Bang in the endless chain reaction through which any particular part of the universe is reborn. In which case, there may well be another Big Bang on its way, one quietly waiting its due time.

The universe is held together by gravity, pulling matter together. More than ninety percent of the content of the universe, however, is unaccounted for. Visible matter, such as the stars, galaxies, planets, nebulae and intergalactic gases, account for less than five percent of its total substance. Invisible matter predominates. In part this includes so-called dark matter, which only interacts with ordinary matter through gravity and is believed to comprise exotic particle relics from the Big Bang. Cosmologists tell us that some mysterious 'dark energy' is latent in space in the absence of matter. They call this *vacuum* energy, and it is making the universe expand at an ever faster rate. The universe is estimated to be composed of roughly seventy percent dark energy, about twenty-five percent dark matter, and about five percent ordinary (visible) matter. Dark matter is, by definition, not visible, since it does not emit, reflect or absorb light. It acts by exerting a gravitational force on real objects. Its gravitational pull is thought to hold galaxies and clusters of galaxies together, while dark energy propels a faster expansion rate that threatens to blow the universe apart. Our most sophisticated equipment cannot detect gravitational waves. The accepted theory of gravity is an incomplete idea that submits to faith on cosmological scales. If we do not yet have the ability to prove or disprove a theory, the honest answer is simply that we do not know. It becomes an article of faith, not a known fact.

But at least there are regularities that can be observed in the behaviour of matter, fundamental regularities that are at work on both macro and micro scales. For example, Newtonian mechanics provides a representation of the gravitational attraction that drives the motion of the stars and planets. Electromagnetic laws regulate the speed of light and hold atoms and molecules together. We can measure the forces responsible for the radioactive processes powering the sun, and the forces at work in interactions of elementary particles. But such regularity, such predictability that allows us to call the phenomenon a law of nature, breaks down with behaviour of particles at the subatomic level. So-called 'laws' become bent, as if the cosmos of our descriptions may not be as thoroughly logical or consistent as we might expect.

The consistency of the natural world, if there is consistency, presupposes the same simple and accessible laws at every step and on every scale, applying invariance to all things, from the smallest events to the most stupendous.

The world is fundamentally a function of the small-scale movements of particles of matter and energy that make up the large-scale properties of material objects. Quantum physics is a description of the behaviour of elementary particles on the subatomic scale, for example the binding of protons and neutrons in a nucleus. In the structure of atoms the nucleus is encased in rings formed by electrons. Quantum systems, however, behave differently from the objective reality of the everyday world of trees and tables. Regardless of how apart they are, particles appear to influence each other by instantaneous communication in an exchange of energy. An object can appear to be in many places at one time. Quantum mechanics alerts our senses hazily to an ethereal whole, out of which the visible universe has precipitated.

According to the Big Bang theory, the universe is expanding and cooling. Cosmic microwave background radiation was first detected in 1965. It is essentially an echo of gamma radiation from some half a million years after the Big Bang. At this stage the universe was cool enough for energy and matter to unblend. The radiation is an indication that the universe was once very hot and dense. In the first millionth of a second of its existence, the universe would have been like a glowing body emitting radiation in a spectrum of wavelengths determined by the temperature. In the nineteenth century the Austrian mathematician Christian Doppler (1803–1853) showed that objects moving towards or away from an observer change both in colour and pitch (we can hear this change in the tone of an ambulance siren as it sweeps past us). The American astronomer Edwin Hubble (1889–1953) in 1929 recognized that the universe is expanding in all directions and that the galaxies are receding from each other. The light from far-off galaxies stretches into longer wavelengths at the red end of the spectrum (red-shifts), so that the greater the distance from the earth, the greater the red shift is. We observe galaxies receding faster from us the further away they are. This increase in red shift, once attributed to an 'escape velocity' in stars, may simply happen because light loses energy during travel, as it gets scattered or blocked by intergalactic dust, with red light energy being weaker than that of blue light.

The corollary to an expanding universe is that it was once smaller. If the escape velocity is known, the clock can be wound back to a point when the universe was but a tiny space bubble, the size of a golf ball or a pin-head of immense density. Before this state of indefinable density there was absolute *nothingness*; at least this is what the finest minds in physics and cosmology tell us. Inside it, a temperature of a magnitude that is utterly incomprehensible to us. Before the Big Bang, there was no space, no time, merely a 'vacuum state', a mysterious energy froth of virtual particles flashing in and out existence in fluctuating electric and magnetic fields. Bursts of high temperatures and energy densities gave rise to a super-heated plasma – the primordial fireball. Bang – our universe was created! Space and time expanded as energy was released as mass; electrons, atoms and finally stars and galaxies were formed in the cosmic expansion.

The picture is almost as satisfying to us as any creation story must have been to our forebears. But for the deeper thinker it has one insoluble problem: in an observed world of temporal transmutations, one is obliged to posit a transcendental reality of energy. If all that exists makes up a quantum state of the universe, in which causality collapses, we will have no empirical handle on any relationship of precedence, unless we can perceive reality beyond evolution and dissolution of gross matter. The Big Bang picture ignores, or at least blurs, the reality it purports to illuminate.

The structure of the universe is underpinned by constants that fine tune the forces of nature. A small alteration in the properties of the fundamental constants would almost certainly cause the universe to collapse. Gravity is the weakest of all of the forces acting in the cosmos, at some forty orders of magnitude weaker than electromagnetism, but it is cosmological in effect. Every object attracts *every* other object in the universe. The further apart they are, the smaller the gravitational force between them. When we jump, the attraction of all the atoms of the earth to all the atoms in our bodies brings us down to earth. Tides are caused on earth by the moon's gravitational attraction on the oceans. The attractive force of gravity causes curvature in the geometry of space and time (space–time) in the proximity of large bodies, analogous to how a cushion or mattress sags with a heavy weight. Stars shine from internal gravitation due to the compression of their huge masses of primarily hydrogen gas under their own weight and the heat consequently generated from the resulting high temperatures. Energy-releasing thermonuclear fusion reactions going on inside stars counter gravitational collapse and cool the stars down permanently in a balanced process that enables a long life-cycle. A tiny increase in the gravitational constant would speed combustion of the hydrogen supplies inside stars, causing stars including our sun to burn out more rapidly. The slightest decrease in the pull of gravity would slow down combustion and chill our sun down.

Light, an electromagnetic radiation, is both particle and wave. Electrically charged particles such as electrons and protons emit electromagnetic waves in motion, hence they are known as 'electromagnetic radiation'. A light source releases energy in little bundles called quanta, in a stream of discrete packets called photons. The strength of radiation corresponds to their frequency of vibration. Electromagnetic force arises as particles exchange energy of light and other electromagnetic radiation carried in photons.

When we heat something it emits electromagnetic radiation. Energy is released from its vibrating atoms. An incandescent object emits energy in a spectrum of colours or frequencies, depending on the temperature it is at. At higher temperatures, more radiation of a higher frequency is emitted. The spectrum of radiation covers visible light, gamma rays, X-rays, ultraviolet and infrared waves, and low frequency waves such as microwaves and radio waves. A slight change in the basic value of the quantum energy constant would riddle the universe with the uncertainty of material form. If the value were just a little more, for instance, fleeting 'virtual' particles, that continually form and disappear in the dark energy vacuum of space, would transfer insufficient energy to transmute into real particles. From the 'vacuum' before time, things would pop in and out of existence. A lower value still, and the atom would be eliminated altogether. Bang would go atomic physics.

Physics needs constants to underscore the way the world works. The speed of light is a fundamental value in the universe. Matter precipitates from energy corresponding to a frequency slower than the speed of light. The smallest change in the speed of light would jeopardize the stability of the universe. A decrease would cause distortions to gravitational pull and upset the process of gaseous collapse in the formation of stars. A tiny increase in the speed of light would plunge the universe into catastrophe within an excessively contracted space and time.

Seed the world

The atomic bomb jolted us to the realization of the interchangeable nature of mass and energy. It threw up the hypothesis of the birth of the universe in a Big Bang from a quantum void bubbling with virtual particles that bump in and out of fleeting manifestations of physical reality.

The idea of a domain of undetected, vanishing energy that envelops the universe dates back to antiquity. The 'vacuum' described in modern physics, however, is unlike antiquity's unchanging void; it is not empty, but a vibrant gravitational domain bubbling with fine matter and energy, from which fundamental particles emerge as excitations and into which they again disappear. Fundamental particles, which are quantum fields of energy, combine in the seething mass of the atom and make up our quotidian world.

As we have seen, the discovery of quantum entanglements of matter and energy has reshaped our picture of reality. There exists more than just the physical that is perceptible to our mind. Objects can no longer be identified from cause-and-effect relationships of discrete parts alone, but are shaped in their properties by the influence of distant entities, all parts in communication with all other parts by virtue of energy streams spread out in space and time. Reality does not unfold as a nexus of discrete properties of disconnected parts. Rather, everything is interconnected, so that events in one part affect another part, and ultimately also the whole.

In our quest to understand the universe, we have been obliged to move from the discovery of 'laws' to the setting up of models. In quantum cosmology, events can happen spontaneously. The hidden 'vacuum state' could spontaneously create particles, and the cosmological expansion of a Big Bang could lead to the emergence of life. Cosmologists speculate that other universes may exist outside our knowledge, that a new universe or part of one could be spawned from a black hole. Nothing of this can be verified by science at present.

Almost anything is possible in quantum interactions, but some things are more probable than others. Quantum systems are possibilities spread out in time and space that can be realized in certain conditions. 'Dark energy' from the vacuum might generate regularities that are favourable to the creation of a universe. Tiny universes might emerge, most of which collapse and cease to exist, but a few could reach a critical size and expand in a Big Bang, going on to evolve into a universe or part of a universe that may be able to support life and beings like ourselves. The prescient 'vacuum' reflects the whole in each of its parts.

Indications are that the mysterious 'dark energy' is accelerating the expansion of the cosmos. The rate of expansion of the universe is currently accelerating and galaxies are moving apart faster than previously – so fast that we or our descendants may not be able to see distant galaxies anymore. This suggests that the universe will continue to grow indefinitely, expanding forever, and will consist finally of bizarre, frozen bodies, remote from one another, floating in empty expanses.

It is becoming more and more difficult to harmonize observation data, so there is currently a proliferation of explanations on offer. What came before the Big Bang becomes anybody's guess. Hence mysticism creeps back into physics. It is no longer as though scientific understanding had eroded religious superstition, but reason and superstition appear to have made an accommodation for it. This has resulted in the concept of a self-explaining universe that arose by chance, with particles combining, genes assembling, messages being carried, cells replicating and so on, all with no particular purpose in mind, as it were, and we as human beings at the end of the chain of development that could just as well have produced something totally different to ourselves, with no finer meaning to live for than our mundane pursuits. Descriptions of quantum mechanics stop at the boundaries of the coarse material world, and give little insight into the finer existence that may lie beyond it. Yet the theory has a metaphysical arm; it embraces everything there is, throwing wide open the locked gates of materialism, and providing a handle for spiritual values like free will, intuition and other stirrings of our inner life. Quantum theory hints at vistas of truth and knowledge that are more universal and authentic than the materialistic viewpoint so prevalent today. Questions of fate and free will are not scientific questions, because they do not originate in the material world, which is the domain of science, but they are and have always been the most vital ones to man.

Whether or not we think that some over-arching intelligence fine tuned the forces of nature and thus enabled consciousness to emerge, or that universes arising by chance in nature succeeded or failed on their own terms, it is a fact that our intellect – a property of the brain and lacking intuitive vibrations – cannot comprehend the primeval 'void' before the evolution of the physical world. Therefore what was before the Big Bang will always remain an enigma to it. But human consciousness transcends the physical, isolating mind from matter. As a potter shapes his pot, so volition shapes human thought and action. The potter's wheel operates from volition using natural forces. From gross matter, something fine is shaped, courtesy of the natural laws. A man cannot *will* without *forming* at the same time, and he has responsibility for everything to which he gives form.

Still song of time

The birth of the earth, some four and a half billion years ago, shortly after the sun had coalesced from galactic dust, stemmed from the gravitational collapse of clouds of hydrogen and helium that subsequently generated stars, which also contained heavier elements like carbon, nitrogen, and oxygen. Clouds of gas and dust are scattered throughout the universe, and condensation of those dust clouds builds stars, like our sun, encircled by planets. From stars that exploded after exhausting all their available fuel, the elements were flung out into interstellar space and formed clouds in which simple molecules such as water, carbon dioxide and simple hydrocarbons were generated. The clouds collapsed to form new generations of stars and solar systems, the planets condensed from rock, dust and gas circling the sun. The earth was formed at a distance from the sun that enabled the primitive conditions for the emergence of life.

Plants occupied the land mass some four hundred million years ago, evolving from simple forms like fungi and algae to stable coalitions of lichens and mosses, which then led to the evolution of ferns and seed plants. But life in some form preceded the first plants. Early life was naturally microscopic. The oldest known microfossils, marine primitive microorganisms, date back to about three and a half billion years ago. The photosynthesizing microbes without a nucleus were the early dominant life form, but are now almost extinct. They enriched the earth's atmosphere with oxygen. That paved the way for nucleated cells, and multicellular organisms such as algae, which have been dated to two and a half billion years ago. These simple microorganisms and other invertebrates in the evolutionary transition six hundred million years ago ushered in more complex life forms, such as vertebrates, fish, reptiles, amphibians and primates. The plants and animals we know so well today evolved in the last five hundred and fifty million years.

Today, our planet teems with a huge biodiversity, with every habitat and every niche inhabited by a wide variety of species.

Habitats transform from one ecosystem to another. A plethora of creatures, germs and microorganisms thrive in every crevice and drop of water in stream, river and sea, in the air and atmosphere, on mountain peaks and in the darkest caves and ocean depths, in rainforests, deserts and arctic wastes. These creatures live in symbiotic relationships, often in the most seemingly unlikely and least hospitable environments.

Each intricate ecosystem is part of a larger symmetry. Species find their ecological niches and each niche becomes inhabited. We may well wonder if such a furthering principle is unique to planet earth, whether our carbon-based life is the great exception in our universe. If life in its ordered complexity is a chance happening of cosmic radiation, then the creation of life, far from being predictable, would be a most unlikely development.

Let us take 'lucky' changes as a possible explanation for life. A species survives by innovating traits and attributes over generations. It 'improves' or adapts – at the expense of others, naturally, for survival – at the species (rather than the individual) level. But we can hardly attribute the sciences, the arts, moral philosophy and all our libraries, cathedrals, temples and palaces simply to a blundering need to survive. These culturally valued artefacts and achievements mean far more than fragile crusts grown on a rational ape subject to a blind destiny.

If species survive only by 'improving' the breed, it would be morally justifiable for man to give nature a helping hand. Detrimental genes could be eliminated and individuals who were likely to add inferior or substandard stock – well, they too could be put, as it were, out of harm's way in order to ensure the success of prime stock, in the manner of pig or cattle breeding.

Darwin never went this far despite an earlier belief that in the struggle for existence, humans should increase their reproductive rate. There must be free competition for all people, he thought, even though many ills attended it. Later, seeing the inhuman conditions that such intemperate competition was bound to create, he modified this view and went on to add: 'Important as the struggle for existence has been and even still is, yet as far as the highest part of man's nature is concerned there are other agencies more important. For the moral qualities are advanced, either directly or indirectly, much more through the effects of habit, the reasoning powers, instruction, religion etc., than through natural selection' (*The Descent of Man,* 1871).

The organized demonization, hatred and the systematic elimination of ethnic minorities resonate through history. But then every survival strategy has its down side. For all that, the use of such powers to oppress the weak would seem to have needed some justification. Eugenics advocates the use of practices aimed at improving the qualities of the human stock, such as encouraging or permitting reproduction of only those people with hereditary characteristics deemed to be noble. At the close of the nineteenth century, it was on hand to justify the repression of 'lesser man' by 'greater man' as a social movement to improve the human species, who could be categorized and selectively bred like cattle and livestock. With credulous schemes of our biological nature hitched to ideology of a pre-ordained destiny, restrictive eugenics sprouted as a social movement worldwide, and included state encouragement of selective breeding and forced abortion, sterilization and euthanasia. White rule, Aryan supremacy and sundry nationalisms that, in their muddled logic, assume one race's superior inheritance over another, led to (and, of course, still lead to) the repression and misery of minorities. From such eugenic seeds do mighty empires grow. And of course eugenics is often used to justify the self-interest of rulers and their cronies.

Once free will or choice is linked to the idea of destiny or 'natural law', men shrug off responsibility and become less than what we call 'human' or 'humane'. In that case, free will becomes an illusion and people trying to exercise it become 'traitors' who must be ruthlessly eliminated to 'save' the populace from their corruptive influence.

In the twentieth century annals, racial theories that were overplayed from eugenic conjectures on human nature, transfixed nations and visited vitriolic savagery upon groups and minorities. Yet the systematic elimination of 'weaker' gene carriers has failed to change the gene pool. In Nazi Germany, the killing of a Jewish person was equivalent to the extermination of a rat; to many a Japanese a Chinese person was considered as inferior, a natural slave fit only to serve their master race and be worked to death. This misapplication of eugenic principles persists, even in the wake of those policies that led to the establishment of gas chambers. In the Soviet Union, India and China there has been government encouragement of enforced sterilization, involuntary euthanasia and attempts to obliterate the 'unhealthy' or 'unfit'. The death knell for the outsider rang out in South-east Asia, Africa and Central Europe from the rigours of 'ethnic cleansing', a euphemism for cultural genocide of people presumed to be racially inferior or in some way different to the dominant cultural or ethnic group.

In any population, an optimal genome – if such a thing were to exist – must devolve on one individual or on no-one at all. If intelligence is our criterion for breeding, from the full set of chromosomes, which ones are responsible for that much-prized faculty? Is it separable from memory and perception? The Pandora's Box of eugenics has been forced open and we will need all our human resources to control it, and to use it for the benefit of humanity and the planet at large.

Line of horizon dimming

Do people commit a crime or offend because of flawed genes that predispose to delinquent behaviour? Our brains translate our thoughts into recordable electrical patterns of neural activity, but not so that we can study patterns from brain imaging scans in order to read thoughts. Do people express violent and sadistic impulses and behave in revolting and unpredictable ways because their brains work abnormally, because crossed nerves have caused subtle changes to the brain's hard wiring, or some neurochemical imbalance has disrupted impulse control? Is the psychopath predisposed to inexplicable surges of serotonin (an important mood-altering chemical) or dopamine, adrenalin or some other chemical messenger in the brain to cause frenzied outbursts of unreasonable anguish and impulses beyond conscious control? Possibly so, but research usually cannot establish a simple association between biology and criminality, nor can it demonstrate anything characteristically abnormal in the brains of violent criminals.

There is wide acceptance that genes have been moulding our bodies since the dawn of time, and that they continue to mould our minds. In common parlance: 'Heredity rules OK?' If this is the case, cultivation of the intellect will make little impression on the individual's cognitive predisposition. In effect, everything we do could then be predetermined and the result of processes beyond our control. If our mental capacities devolve primarily upon our biological nature, then we – whatever the 'we' represents in these processes – are responsible neither for our sublime thoughts nor for our dastardly actions. Virtue and vice, if such terms are to retain any meaning at all in these circumstances, are therefore merely artefacts of biology.

Can a criminal be reformed? Is there even such a thing as criminality? Can we be ultimately responsible for our actions? We are, of course, capable of scrutinizing our own motivations; we use our ability to think to create ends of our own, and we have the urge to claim our own life. Otherwise, who or what is to restrain the individual from anti-social behaviour? Who is to decide who must be restrained and who goes free, who deserves life or death? With the spread of national databanks of genetic fingerprints, the implications for civil liberties are horrendous.

As health-care costs continue to mount, with growing dependence on ever-more sophisticated technology and an expanding elderly population, the question of who 'deserves' treatment and who does not also arises. What of chronically disabled people or those who will need a lifetime of expensive care? What of the dying and the terminally ill? Is there ever a case to justify voluntary or involuntary euthanasia? How can we justify the often enormous expenditure involved in prolonging the life of a patient (frequently these days, and significantly and cynically termed the 'client' or the 'customer') who is either comatose or whose every breath is both an agony to himself and to his relatives?

Temptations abound for the economist, the politician and the genetic engineer eager to employ the most up-to-date technology, to make yet another scientific 'breakthrough'. Perhaps it is fortunate that the human genetic machinery is so complex that it may prove impossible to tamper with the gene pool and 'improve' our breed. Otherwise who knows what might happen if the tools for human genetic engineering become available? Splitting the gene would wreak even more global havoc than splitting the atom has.

Each child comes into the world inheriting nothing of the personality of the parents. The child brings along its own identity and the foundation of its self-confidence, and has an inner potential, something incalculable from way back, which will be developed in unknown ways through pressures from its external and internal environments in the course of its life experiences. The child may inherit physical qualities or dispositions from the parents, who may nurture his moral and intellectual development, but the personality is not inherited, and the parents will have little effect on the personality that emerges. This is simply his and his alone. Spiritual inheritance is a myth. Though a child may be guided spiritually by the parent, the latter cannot part with a bit of his or her own living spirit.

There are spiritual just as there are physical affinities between people, and this includes family members. But while physical characteristics depend on genetic inheritance, our spirit core of being is attracted to the circumstances of birth through a finer process. According to the work *In the Light of Truth: The Grail Message* by Abd-ru-shin (Oskar Ernst Bernhardt, 1875–1941), a mutual longing of homogeneous beings to come together is a consequence of the furthering principle of a cosmic law – the Law of the Attraction of Homogeneous Species. Between human beings, homogeneity may be dictated by unfolding abilities, special talents and gifts, and the individual's nature and character as well as the general direction of striving; but it may also be dictated by weaknesses, character flaws and propensities. The prevailing characteristics radiate respective hues or frequencies, which transcend the earthly domain. The characteristics that correspond most closely between mainly a mother-to-be and (to a lesser degree) a prospective father and a soul, a discarnate human being in the beyond, form the basis of an attraction that may lead to the soul's incarnation at some point through pregnancy. Between parents and their children, heredity is limited to physical traits; any similarity between their character traits, artistic abilities, and so on, is dictated by spiritual attraction of homogeneous species.

Beyond the cell

We shape our own reality by the nature of our volition and striving. We think, create our own values, even in accepting the beliefs of others, and we choose ends of our own. Our physical body is affected by our volition. Our astral body, an inner body of finer material substance, is a template or prototype for our outer physical body, but at the same time it can be enhanced or disfigured by our thoughts during the course of our earth-life. Thoughts acquire energetic forms in the same consistency as our astral body and have a moulding effect on this body. As a thought presses toward manifest realization, and is reinforced by similar thought-forms through the attraction of homogeneity, the physical body, being of a much coarser consistency, does not change to the same degree as the astral body during this process (except, perhaps, something about the facial expression, which can reveal much about our character, especially as we get older). In this way, our thoughts have a bearing on ourselves and on others.

We are not all born equal. Many are born poor or disabled, seemingly born to suffer, certainly born to die. We all must die one day, no matter what differences may exist between us. But we all have the power of volition, the ability to exercise will consciously. We are partly self-determined. Our view of the world is shaped by our own ideas, informed or otherwise, and we bind ourselves in the self-formed shaft of our own thoughts, our desires, hopes and wishes, our beliefs and convictions.

Without free will we would have no need for words like 'good' and 'bad'. Yet these words have been needed, as far as we know, from the earliest days of primeval man. And we understand them in an absolute sense, as meaning something more than 'good for me' or 'good for the human race'; one man's 'good' is not another's 'bad'. Our perceptions change as society changes over time, but there is a difference between saying one man's 'good' is another's 'bad' and saying that our minds may provide us with an ability to grow in an as yet unrealized self-affirmation of the never-changing cosmic order. Human ennoblement ensues mainly from expansion of our self-awareness. We have a deep moral awareness to guide us in everyday life. Even though we realize that we frequently fall short of our ethical aspirations, and that 'good volition' fails miserably on occasions, we do strive for ideals. These ideals cannot be discovered in the laws of nature, but each one of us feels them within our self. They seem to come from somewhere beyond imagination, from intuition, from a sense of being subject to a higher law or principle, something beyond the shackles of time, a reality that existed long before our universe came into being.

The human spirit is on a passage to awaken to a self-aware personality. We start out as an unconscious spirit germ or seed-grain from our plane of origin (the spiritual world lying far above the gross material world) harbouring an intrinsic urge to develop consciousness. The journey leads us as far away as down to the earthly domain, acquiring a bodily covering at each transitional plane along the passage. We appear on earth in a physical body.

Our developing spirit possesses our dominant will. It distinguishes us from other creatures, and it is able to liberate us from domination by instincts. By its exertion, we stand to transcend our earthly bonds.

We are creative. We can combine existing elements from the natural world in various ways. From the flint arrowhead to laser technology, our technical know-how advances by acts of will, by understanding how the power of our inner perception can work on the material world. Let us consider a journey that, for the sake of simplicity, takes us to that proverbial 'desert island', a pristine locale where no human has set foot before. Everything we discover is already there. If we gather and stack stones to build a camp-fire site, we add a fireplace to the island. This is an action of technology, an adaptation of the forces of nature. In its noblest consequence, the creative act stems from intellect that is mentally directed by spirit working on matter manifestly in accordance with the cosmic laws, which are an expression of God's will.

Technology is an artefact of the intellect. Thinking, too, is an action of the intellect. The intellect issues from the brain. It is the executive instrument of spirit, and it implements the volition of our spirit into the earthly. It is *not* the spirit itself. Control of the body belongs to the spirit. If today the intellect is seen as the spirit, it is one of the great errors of our time. The spirit must liberate itself from the bound will of the intellect. Bound, because being a product of the earthly brain, the intellect will never be able to transcend the earthly, and neither can a spirit that has enslaved itself to the intellect comprehend anything beyond earthly conceptions of time and space, unless it is able to break this shackle from within.

True, it is by the intellect that we begin to sort the boundless sensory information we process. We receive impressions, learn to form and associate ideas, and communicate ideas with words. Thoughts and words are acts of will pressed by spiritual power. As such they carry responsibility. Word-making is a distinctly human activity and words can weave an enormous power. Language can enlighten, bewitch or bamboozle us. It is the most patent and potent aspect of technology, but it is really much more than that, so that its misuse can have dire consequences for individuals, as well as for entire nations. How apt is the expression that 'words come back to haunt us'!

We rely on the various branches of technology every day, but we need to remain in control of how that technology is applied, for to allow it to monopolize our thinking would be detrimental to us and to the environment. The atom bomb, germ warfare, and the pillage and rape of the planet's finite resources – the fiery signs of the heedless uses of technology – are everywhere, from acid rain to flood and famine.

We are culpable for the unwholesome state of our technology. Our motivation is open to choice, our activity working voluntarily through the agency of free will. The state of the planet and the ecologically disastrous future into which we are plunging it through our ill-conceived attempts to gain the upper hand on nature, or to tame its raw power or brand our cultural trademark on the landscape everywhere, is clear testament to the unwise use of our capacity for free will. We remain ignorantly indifferent, despite the strong indications that much of our cultural behaviour, which might have given our forebears a survival edge, today threatens disaster.

In reality nothing is invented – only found. The scientist or artist does not actually create anything new but rather discovers what already exists, waiting to be uncovered, or realized. Deep inside us lies a picture, a stark reflection of finer realms and ideals taken in by the spirit germ in the course of the passage down from our origin. Its principle of beauty anchored deep in us is the simplicity of the omnipotent will manifest in the cosmic laws and forces of nature, and it has from the outset animated our urge to moral virtue. This apprehension of moral beauty has always existed as an objective value of our yearning for higher values and our inner adjustment, and remains no less so in a diminished age of scepticism. Beauty reminds us of the eternal. In its melody swings purity of truth, whether in the beauty of nature, the ideal of beauty, or art. However ingenious our 'inventions' might be, they tap into an enduring neutral power of the cosmos that we may *will* to good or ill expression.

Not infrequently, the latest 'discoveries' turn out to be makeovers of artefacts belonging to a bygone age. Key ideas and objects spring forth in different places. Great cultures have declined, and technologies perished, only to be rediscovered in later cultures. Modern man is rooted in his past, and those past images refresh our perspective in the present.

Like the seasons and the motions of planets around their star, some eternal cyclic principle appears to drive the universe, whereby that which has passed away will return renewed and strengthened, offering a new impetus for development. Thus spring will follow winter, bringing a new awakening, and the death of stars or galaxies is followed by the birth of new ones. Despite our enthusiasm over the latest discoveries about the universe and our pride in our brand new scientific research tools, things happen again and again – with refinements and modifications, naturally – but affirming the Will that rules the universe, which is expressed in eternally unchanging laws that we cannot circumvent or overthrow.

Worlds evolve and disintegrate in the regularities of nature. The shaping spirit is active everywhere, in the tiniest processes on earth to the most stupendous events across the universe. Men die, species die out, cruelties and injustices on occasions proliferate, and then they are purged; lessons are learnt and then forgotten. History repeats itself.

Lives seem precarious and expendable, but life itself is precious, a given for which we cannot quite account. We regard it as self-evident that nature is guided by laws, and these are laws to which we are also subject. And if a law exists, so too must a Law Giver. Some people deny this of course, maintaining that everything is driven by chance. But even chance itself is governed by laws, the details of which may as yet be unknown, otherwise we would not be able to evaluate random events using statistical methods, and it is also totally illogical to assume that order evolved from chaos. Order must have been present in the first place, which presupposes an all-pervading will or power that drives everything towards perfection within its species and automatically casts off any rotten fruits. It matters not whether we call the power behind this law God, or the Great Spirit of the universe, or any of the ninety-nine Islamic kennings for Allah. It merely signals the metaphysical limitations of a developing consciousness. Recognition leads towards a God who impressed laws that determine the possibilities for development. Power from God pulsates through all that exists, and translates into energy that moves matter and thought. It must be said here, though, that contrary to the views of the pantheists (who deny that God is 'totally other' than the cosmos or that God's Being is distinct from it), or of those who seek God within themselves, God Himself resides outside Creation and does not merge with it; it is only in His Will, in the form of His Laws, that we recognize Him throughout Creation. *In the Light of Truth: The Grail Message* by Abd-ru-shin likens the relationship of God and His Creation to that of an artist and his work. The artist becomes familiar to us through his work. We do not have to know him personally. For instance, musicologists who study Mozart's musical output will probably know more about the 'real' Mozart than his contemporaries ever did, even if they were counted among his friends and acquaintances. In the same way we can come to the recognition of God by studying His Creation and its laws, and this makes Him appear omnipresent to us. However, in spite of this, by the nature of our species as spirit, and the insuperable gulf between the species, we are and we must forever remain at an incomprehensibly vast distance from God and from the divine.

The natural world is the outcome of sustained development. Life evolved on earth from the simplest beginnings. The biochemistry of our bodies is imprinted with the legacy of the trails of biological evolution that lead back to a few common ancestors with the rest of life on the planet. Fossil record testifies to intense upheavals in the evolution of the planet. Milestone relics, such as huge coal seams, dot the landscape. Under the ground and under the sea we have discovered vast deposits of oil and gas, fossil fuels that were once vast forests but became transformed by nature into tools for survival, to the extent we can use them. The drastic upheavals have been turning points in the development of the natural world. Disasters convulsed whole regions, caused mass extinctions and wiped out species. At the same time, like tidal waves, the pulse of development moulded conditions for species of plants and animals to emerge, with new landscapes and ecosystems, occasionally in a frenzy of innovation. We now know that we are linked to all these species.

There are more living species today than ever before, far more than a billion or even a million years ago. The world has grown from a very simple state to become more highly organized and complex. Not every single step of that development was fixed from the very beginning. The details of all the diverse life forms were not predetermined to the *n*th degree. The realm of possibilities was in place, the great whole and pattern determined by the unchanging cosmic laws. The possibilities accommodate the freedom of will, which arrives with our spirit species and brings about the transformation of a whole artistic, spiritual, cultural, and intellectual world. The exact size of every star, the structure of every planet and of every life form was hardly determined at the outset. Life in some form probably existed on one or more stars or satellites long before earth yielded its tiniest microorganisms.

Giant steps

The picture of our origins is blurred. In Africa some seven million years ago, it is thought, the last common ancestor of humans and the apes diversified in separate primate lineages. While the exact nature and living conditions of this common ancestor are uncertain, fossil fragments of ape jaws seem to bear out the diversification. The earth then was mostly warmer and wetter, with large parts covered in lush tropical forest. Apes in an enormous variety of lineages dominated the landscape. Five million years later only a few modern species, such as the orangutan, gorilla and chimpanzee survive. The earth has become colder, and the forests have dwindled and become grassland. There is mass extinction in the ranks of all land species from changes in selective pressures such as diet and climate; creatures that were adapted to forest habitats are being wiped out. Fossil analyses suggest that early humans may have coexisted alongside the australopithecine hominids three to four million years ago.

The new environment favoured species that could adapt to the open plains, amid high competition for food and formidable predators. Many plains-dwelling species emerged over the few million years that followed. Later, too, our upright-walking species would appear, one that would be enabled to high attainment because of its capacity for culture and language.

The first major step in the evolutionary divergence of humans occurred with the emergence, roughly seven million years ago, of a bipedal chimpanzee-like species from a quadrupedal, ape-like ancestor. This was the dawn of the hominids, when creatures with hands and feet that became more closely related to humans than to apes began to walk on two legs, and moved out from the forest floor to the plains.

Then, some three to five million years ago, hominids appeared, that walked erect (although they looked more ape-like than human). They had a smaller brain and a protruding jaw. Their four limbs – with shorter legs and longer arms – were adapted for grasping trees. These hominids (misnamed as man-apes) used stone tools to obtain food as a first step on the long road to the use of technology by humans. They formed the divergent lineage towards modern humans.

About two million years ago a more human-like upright creature lived on as a survivor from a crouching ape-like hominid. It could use fire for cooking, for warmth and for fending off wild animals; it could hunt skilfully, run like a modern human, make stone tools and weapons, and make clothing. Its appearance was already so human-like that it is considered today to be the first true human. This hominid was the culmination of a giant climb atop the evolutionary pyramid. The species would eventually migrate beyond the African landscape and spread to all parts of the world!

Today, following in Darwin's steps, we are inclined to see this material world that we experience as the only reality there is, and as the basis of our morals and values. Science is a part of culture, but it is as though by its rationalist limitations it has enveloped all our finer feelings. We become another outcrop of coincidence in the game of chance that is evolution. We are left with no over-arching direction for our development, with no higher principle behind the expressive intensity of culture, or our attributes of language, art and music. Yet it is from the ideas of modern physics in the last century, which have recast conceptions of the material form, that intimations sprout about the working of an omnipotent Creator.

Independently of Darwin, the Welsh naturalist Alfred Russel Wallace (1823–1913) arrived at similar ideas of natural selection in evolution – both their books were published in 1859. However, they differed on how natural selection applies to man. Wallace regarded some capabilities, for example self-consciousness and musical gifts, as no mere adaptations shaped by selection. Darwin insisted that selective pressures had shaped the human mind as a new adaptation, and 'the difference in mind between man and the higher animals, great as it is, is certainly one of degree and not of kind' (*The Descent of Man,* 1871). Wallace was certain that some other cause beyond selection based on environmental differences was responsible for human self-consciousness, lifting us above the biological determinism that is very prominent in animal life. He contended that to 'prove continuity and the progressive development of the intellectual and moral faculties from animals to man is not the same as proving that these faculties have been developed by natural selection' (*Darwinism,* 1889).

Wallace envisaged two stages in the evolution of man. In the first, bodily development followed the same evolutionary path of natural selection as all other species, with the organism adapting to the pressures of the environment. In the second stage, our will emerged as an attribute. It was able to influence natural selection and shape the environment to our interests, and in the process modify our physical form. Faced with 'the higher intellectual and spiritual nature of man',

Wallace asked why our distant ancestors had evolved brains that could devise language, compose symphonies and do mathematics if *all* they needed for survival was a gorilla's intelligence. In other words, the early hominids evolved thinking strategies that helped them gather fruit, evade predators, survive, and reproduce. Wallace states in his book *Darwinism* of 1889 that ‘even when carried out to its extreme logical conclusion’ Darwinism not only opposes ‘but lends a decided support to a belief in the spiritual nature of man’. He explains that the human body ‘may have been developed from that of a lower animal form under the law of natural selection’ and goes on to say that ‘we possess intellectual and moral faculties which could not have been so developed’. This aspect of human development he ascribes to another origin, stating: ‘for this origin we can only find an adequate cause in the unseen universe of Spirit’.

He buttresses his argument in his *Contributions to the Theory of Natural Selection. A Series of Essays* (1870): ‘If a material element, or a combination of a thousand material elements in a molecule, are alike unconscious, it is impossible for us to believe that the mere addition of one, two, or a thousand other material elements to form a more complex molecule, could in any way tend to produce a self-conscious existence. There is no escape from this dilemma – either all matter is conscious, or consciousness is, or pertains to, something distinct from matter, and in the latter case its presence in material forms is a proof of the existence of conscious beings, outside of, and independent of, what we term matter’.

The biblical narrative of the origin of the universe, in which the earth is created in the course of a busy week, is simply one of man's many attempts to explain phenomena that are beyond the grasp of the intellect. The evolutionary theory is another. The story of the creation is more than *either* of these attempts to explain it.

The look of life

Humans perceive, just as we experience sensations, thoughts and emotions in our conscious awareness, but we mostly narrow our attention to the world of phenomena that our senses register, aided by any diverse technological instruments at our disposal. We retain striking experiences in our memory. Most experiences leave no permanent trace, but we may recall a learned skill, a song, a face, an acquaintance's name or an event from our childhood. The limitations of our senses tend to define reality and our world of ideas, although the facts do not speak for themselves. We see what we are prepared to see at any given time, based on old impressions and learned associations that are derived from our unique experiences and memories. We mine the shaft of memory and experience for images to see the present with. We look for evidence of our preconceptions. A few centuries ago, for example, it was customary to argue that a clear drop of water contained no living organism. Powerful microscopes now provide sound evidence that the 'pure' drop is, in fact, teeming with millions of free-living microorganisms locked in a battle for survival. Well into the last century, our understanding of the universe was framed within the bounds of the resolving power of optical instruments.

We seek to supplement or amplify elements of experience such that the perception corresponds to our expectations. Our perceptions are also evaluated in the context of the environment and present influences. We know that the world is not as it appears to us. Each person carries within them a unique image of their surroundings. Language provides uniform symbols to represent much of what we see, hear and perceive, but everyone labels their perceptions in their own way and seeds their motivations and choices. We each experience or learn from a first-person perspective. The entire visual process from the eyes to the conscious awareness of what we see, which can be individually so different, often represents the basis for irreconcilable differences of opinion. 'Two men looked out from prison bars. One saw mud, the other saw stars' (attributed to *Frederick Langbridge* 1849–1923).

Five senses register the world to our experience: seeing (the eyes are sensitive to light), hearing (the ears respond to sound waves), touching, smelling and tasting. The sense organs generate electrical signals. The brain further processes and evaluates the signals from the sense organs, which are in effect the outposts or sensors of the brain.

Other senses include the senses of time and balance (the sense of body awareness). Even exhaustion, hunger, thirst, pain and well-being press in on our consciousness, but they are outside the five senses mentioned. The sense of balance we owe to an auxiliary function of our ears. The sense of time arises because our senses perceive changes that are evaluated in our conscious awareness in relationship to changes in the course of the day or year.

Our perception of reality is closely related to our experience of time. Contradictions arise in the time order of our senses. We illuminate the firmament with our imagination; what we observe in the night sky – the stars twinkling, for example – is the effect of light rays that reach the earth thousands, millions or even billions of years after they left the star. That star or the constellation it was part of may have long since reached the end of its life-cycle and imploded in a gravitational collapse, or even exploded in collision with a comet. We perceive what can be proved *not* to be there. In this instance, appearance and actuality often do not necessarily accord. Or in colloquial terms, things are not always as they seem.

Ever larger radio-telescopes, infrared monitors and space satellites, and techniques for detecting X-rays, gamma rays and the ultraviolet and infrared cosmic radiation bombarding the earth – all these things have dramatically changed our world picture. Yet, as we sweep the heavens with more sophisticated instrumentation and probe the small-scale world of the atom, we encounter constantly expanding horizons to the processes of this world that require never-ending refinements of our technology to explore. We thereby come to realize that we understand but a tiny part of the physical world around us.

The workings of uncontrollable chance play a pivotal role in modern scientific schemes about the development of the cosmos. For example, unpredictable quantum fluctuations of fitful subatomic entities flitting in and out of a primordial void brought the universe that is law-governed into existence. Quantum mechanics deals with the radical rules that prevail on very small scales of the building blocks of matter. And to grasp a handle on this randomness at the heart of matter we are left to couch reality in mathematical arguments or resort to allegories of life, love or consciousness to a significant extent.

Today, we know that matter together with its constituent atoms, which are in turn made up of tiny nuclei bound to electrons by electromagnetic forces, is fundamentally indeterminate; more an illusion than reality on a micro scale. Nuclei consist of quantum entities, which are one hundred million or more times smaller than the atoms, intermingled in systems of protons and neutrons, bound by intense nuclear forces of a short range. In this way, matter built on subatomic quarks and electrons resolves into an impenetrable texture of wave-like particles and particle-like waves. Particles can disappear and turn into pure energy, which, in turn, can be converted back into particles. Like time and space, matter and energy defy definition in the context of their universal intermingling, which was the way things were at the dawning of the universe.

There is an ethereal void of potentialities and probabilities of energetic excitations bathing this world in radiation. From energy exchanges between radiation and matter, wave and particle (that is, the interaction of atoms and molecules with radiation) can condense to actualize everything we sense and operate at the heart of everything in the universe. Nothing is ever without a trace of energy and everything emits radiation in a frequency that corresponds to its consistency; for example, the magnetic field of an electrical current or the electric field of a moving magnet. Electric and magnetic forces are complementary aspects of an energy field; they are changing electric and magnetic fields of electromagnetic waves.

Tempting as it is to say that only material things exist, because our cerebral faculty is matched to only understanding material reality, we are forever witness to the inexplicable, the immeasurable, and the immaterial. For everything we see is governed by what we do *not* see, and the universe accessible to our senses is by no means the only reality there is.

Angels on a pinhead

Quantum mechanics, as we have seen, deals with reality on minute scales of time and distance. It deals not simply with very small systems of matter, but with 'activities' and 'forces' encountered therein. In short, it deals with the unpredictable behaviour of matter encountered on a microscopic scale.

Without the seemingly bizarre laws of quantum entanglements there would be no stable atoms, and the sun would produce no energy. In quantum mechanics, matter behaves as particles and as waves in different instances. An atom, for example, is known to be a particle, and is detectable as having a well-defined position. Thus a particle is usually conceived as point-like. But a wave fills all space. Waves can propagate through any imaginable medium and where no medium is apparent.

It is well known that when a substance is heated it glows. The radiation emitted depends partly on the substance being heated and on the temperature. A lump of iron heated in a fire first glows deep red at moderate temperatures, then orange, then white, and finally blue at very high temperatures. Heat causes the molecules and atoms that make up the substance (themselves ensembles of electrical charges) to vibrate. The oscillating charges emit electromagnetic radiation (heat and light), with the red glow of longer wavelengths appearing at moderate temperatures, and shorter wavelengths appearing at higher temperatures. The shorter the wavelength, the higher the frequency of the wave, that is, its cycles per second.

In general, a body emits radiation at a given temperature and frequency just as well as it absorbs the same radiation. Glass, for example, absorbs little of the light that passes through it, and a metallic surface may simply reflect much of the light that falls on it. A black body is an object that absorbs and re-emits all radiation incident upon it. A hot black body emits radiation more at the higher-frequency, shorter-wavelength part of the spectrum than a cool black body. Max Planck (1858–1947) explored how the energy distribution among different frequencies in the black-body radiation was related to its temperature. This was around 1900. His supposition that black-body radiation is not continuous, but emitted or absorbed in discrete packets of a definite size, would permanently our conception of the world. The clear and determinate process of physical reality devised by Newtonian classical principles was radically overhauled at its subatomic roots by quantum mechanics.

Experimental measurements of the spectrum of radiation emitted by a black body did not fit the predictions of classical physics. Planck found a way out of the dilemma by devising a remarkable formula that related the energy and the frequency distribution of radiation, and this agreed precisely with the available experimental data. It assumed that the oscillating charges of vibrating atoms and molecules could not be radiating or absorbing energy in smooth, continuous waves but only in a punctuated process of discrete energy chunks. The frequency depicted the rate of oscillation; the higher the frequency, the more the corresponding modes of oscillation. Planck uncovered a punctuated behaviour of radiation in which energy was neither lost nor gained in a continuous fashion but only in discrete amounts. Each discrete bundle of energy he dubbed 'quantum' (plural quanta from the Latin *quantus*, meaning 'how much'). The energy content of each one of the quanta was proportional to the oscillating frequency of the radiation.

Light was shown to carry energy bundled in these discrete little packet-like particles or quanta. The radiation quanta of light came to be called photons. Temperature reflects motion in as much as particles move faster at a higher temperature. The size of the energy packet depends on the frequency of oscillation, and colour is an indication of frequency of the oscillating wave. The greater energy emitted at higher frequencies came in larger and larger chunks. Planck's formula expressed this in terms of energy quantization, relating the particle-like property of energy to the wave-like property of frequency.

When an object is heated, such as coal in a fireplace, a light bulb filament, an oven or an electric coil, it emits light energy or photons of different frequencies. In the process it takes on colour and brightness according to the temperature, and the spectrum of electromagnetic radiation depends on that temperature. Objects at room temperature also glow, predominantly at the infrared end of the spectrum, which is outside our visual range.

As we have already seen, an iron bar heated to increasing temperatures glows first red, then white to blue, and brighter still. White light is made up of the full spectrum of electromagnetic radiation, as in a light bulb filament, where electrons in an excited state radiate energy as light. The higher the temperature, the greater the radiation given off at higher frequencies. For Planck, the particles in an atom oscillate like coiled springs, emitting and absorbing photons of energetic radiation. The indivisible quantum value is proportional to the frequency of vibration or the rate of oscillation. An elementary energy quantum characterizes a certain frequency. The higher the frequency, the greater the energy content must be.

In classical physics, nature makes no leaps. Adding more and more energy to a glowing object would lead to unlimited amounts of electromagnetic radiation. There is a minimum energy in an atom at a certain vibration frequency, the amount of energy needed by high-energy electromagnetic ultraviolet waves, x-rays and gamma rays to attain their high frequencies is prohibitive, so these most energetic forms of electromagnetic radiation do not bathe the world. Instead, lesser energy waves predominate, while stratospheric layers shield us from the high-energy emissions of the sun.

Although the rules of quantum mechanics apply to the behaviour of matter on a molecular and atomic scale, the effects upon everyday objects are too minuscule to notice; thus the equations of classical physics can still be applied here. Quantum mechanics nevertheless has implications binding upon the universe at all scales. For example, it provided the microchip technology that opened up the information superhighways; it led to the development of the radio, transistor, television, laser and nuclear power. Computerized axial tomography (CAT) scans and magnetic resonance imaging (MRI) technologies – used for evaluating brain activities – also depend on it. The paradoxes of quantum processes explain the production of nuclear power by stars, sunlight, electric currents and superconductivity, as well as the range of colours produced by chemical compounds and the strength and flexibility of solids.

In his equation Planck made the particle-like property of energy proportional to the wave-like property of frequency. The *constant* of that proportionality is the ratio between a particle's energy and its frequency. Energy from the sun enables life forms to exist and develop on earth and makes possible the interdependencies among the earth's living system.

For example, energy from the sun is absorbed during photosynthesis by the plant biomass, which represents vast amounts of fixed carbon and enormous reserves of water, influencing the planet's climate and atmosphere on a large scale. The sun's surface temperature is roughly five thousand eight hundred degrees Celsius, or eleven thousand degrees Fahrenheit. This gives it a white colour that often, because of atmospheric scattering, appears yellow when seen from the surface of the earth. If Planck's constant were half its value, then at the same surface temperature the sun would no longer appear yellow. Most of its wavelengths would be violet. If the value of the constant were doubled, the sun (at the same temperature) would glow infrared with longer wavelengths. Then we would not be able to see it at all, at least not with the kind of eyes that living things on earth have. We humans only see within a narrow electromagnetic band, limited to a narrow range of wavelengths that covers the visible colours of the rainbow (violet, indigo, blue, green, yellow, orange and red). This range can be widened using special devices like night-vision goggles, which enable infrared vision. The world, together with our own evolution, has depended on the predominance of that narrow spectrum of light.

Movement at the speed of blur

Our insight into the operation of the quantum mechanics that govern the subatomic roots of physical reality was only gained slowly. In 1905, Albert Einstein suggested that a light beam travels as a hail of particles (rather than as a wave) – the same quanta described a few years earlier by Planck. Light is in this sense 'tangible', meaning that it is a wave at some times and a particle at others. In 1912, Neils Bohr made similar assumptions about atoms to those Planck had made about radiation. He determined how electrons formed only certain orbits around the nucleus at the centre of an atom. The electrons can only exist at certain energy levels, or orbits at distances from the centre. All other orbits are precluded. Successive orbital shells contain only a precise number of electrons, and it is the electrons in the outermost shells that are most engaged in chemical reactions. Bohr wondered why electrons whizzing around the nucleus like the planets around the sun did not lose energy. According to classical physics, the electron must radiate away some of its energy as it spirals round the nucleus. What, therefore, makes an atom stable? Why is a table or chair solid? Why do all electrons and atoms not simply collapse to a ground state of the lowest possible energy? The lowest energy level, that of the inner shell, is located at only a slight distance from the nucleus. The idea then developed of electrons jumping between allowable orbits, like rungs on a ladder, emitting or absorbing photons in the process, and this meant that energy could only change in irreducible little steps of quanta. The energy difference between the successive orbits corresponds to the frequency of the photon emitted or absorbed. Atoms thus seemed like miniature solar systems with electrons making quantum leaps in a transition between discrete orbits. But electrons do not really orbit; each electron exists in something more like a fuzzy cloud that can vary in size and shape, according to its energy level, or state. Several different-shaped clouds can exist at the same energy level. Bohr's model of the atom, however, could not explain why the electrons were limited to some orbits only.

Particles fall into two categories. Protons, neutrons and electrons belong in one group, behaving like particles of solid matter (matter particles) and making up the nucleus and atom structure. This particle property of electrons underscores the arrangement of chemical elements in the periodic table (and all chemistry). The second category of particles includes photons, electromagnetic and nuclear forces, which behave like waves and make up the binding forces ('force-carrying particles'). Four fundamental forces of nature acting at distances are known in science: gravity, electromagnetism, and the strong and weak nuclear forces acting from within the nucleus of atoms. These nuclear forces are generated as matter particles exchange force-carrying particles with each other. Any number of each of the second-category particles can combine into a single giant particle. Such is the case with laser light, for example, in which millions of photons band together to resemble one giant photon.

If waves can sometimes behave like particles, why can particles not sometimes behave like waves? This was explored by Louis de Broglie in 1924. If light could behave under some circumstances as particles (photons) and at other times as waves, then matter, too, such as electrons, could have characteristics of both particles and waves. Thus electrons and photons could lead double lives, depending on their circumstances.

There was, however, still the problem of explaining the stability of the atom, why all the electrons in an atom do not just fall into the single orbit of the inner shell, losing their energy in the form of light in the process. The inner shell has only one orbit and can contain just two electrons, while the next shell contains four orbits and can contain eight electrons. In 1925 Wolfgang Pauli proposed that nature must have instituted an exclusion principle so that no orbit could contain more than two electrons, and that no two electrons in an atom could be in the same energy state at the same time. The same applies to all particles of matter. The electrons in an atom occupy different energy states, or levels, spread throughout different orbits. But only certain electron orbits, or energy levels, are possible within the atom. These primary energy levels have gradations or sub-levels. Two electrons that are at the same energy level spin like tiny magnets in opposing directions. The electrons cannot pile on top of each other in the lowest energy orbit. If an energy state is occupied, or an orbit is already fully occupied by two electrons, the next electron takes on a higher energy state in another orbit, further out, so the energy levels are filled up in succession from the lowest energy level of the single orbit in the innermost shell, to the next shell, and so on to the highest, outermost shell. The higher an electron's energy, the farther away it is from the nucleus and the weaker the power of attraction between the two will be. This is what keeps the atom from collapsing to its lowest, ground state. The more electrons that are added, the bigger the atom becomes; the more protons that exist in the nucleus, the more strongly the electrons in the outer ring are attracted. The equilibrium of these forces keeps the atom stable, and is responsible for the existence of solid matter, and ultimately for the diversity of chemical elements and material forms found in our universe.

The familiar model is of an atom with a clutch of electrons whizzing around its nucleus in elliptical orbits, but, as we have seen, this is not correct. As a particle, an electron can be bumped from orbit to orbit; as a wave, its orbital pattern rings the nucleus like a cloud, and under the exclusion principle only a certain number of orbits are permissible. The arrangement of atoms determines the chemical properties of the elements, from the hydrogen atom with one electron in the first orbit (the first element in the periodic table) and helium with two electrons fully occupying the first orbit (thus making it an inert element) and so on to uranium, which is element ninety-two.

Pauli's discovery explained the chemical behaviour and diversity of elements, and it also had far-reaching consequences for solid-state physics. In a crystal, atoms are packed so closely together within the lattice structure that their electrons continually collide. In quartz, for example, electrons vibrate at extremely high frequencies. According to the collective exclusion principle, no two electrons may be found in the same physical state. In solids and metals, the energy levels of the electrons of the constituent atoms shift to produce two bands of energy. In the lower-energy band, the so-called valence band, the electrons are captive, bound to individual atoms. This is due to the tendency of electrons to fill the lowest available energy states. When heated, electrons from this band can absorb energy, and with enough energy they become free to move and leap over the intervening gap to the next, upper conduction band. This makes the material electrically conductive. In the higher-energy conduction band, electrons can move freely within the atomic lattice of the crystal and can conduct electricity, creating an electrical current. The gap between the two bands is the critical characteristic of the electron pattern within the atom. Materials that are electric conductors have overlapping bands, non-conducting isolators have a large gap between bands, and semiconductors have a small gap between bands.

This prohibition of egalitarianism of electrons is not fully understood, but it is interesting to note that attempts to impose egalitarian principles on human society by the advocates of the various forms of socialism and communism have largely failed. We may thus be dealing with a universal principle here, which also has implications for our own species.

Let us explore this idea a little further. Quantum descriptions resonate in true affinity with finer concepts like free will and responsibility. Free will precludes a manifest equality of human experience. The ubiquitous cosmic laws mandate the same opportunities for all, and respond incorrigibly to each volition, thought and action, transporting us to various states (in this case, of a homogeneous spiritual kind) analogous to entities emitting and absorbing energy, whereby each individual attains a different state of experience through his own free will. The exclusion principle, in this context, is a key concept with respect to the diversity of nature. It was also the starting point for the breakthrough technology that led to the construction of the first transistor, and culminated in the development of microchips and microelectronics.

In 1926 Erwin Schrödinger suggested that electrons orbit in a wave undulating in a 'spin' around the atom's nucleus. The wave pattern, in its wholeness, determines not only the size of the orbit but also the distance of the orbit from the nucleus. The wave is an energy level of probable location of the electron. In essence, it is a stationary wave spread out at the energy level. Since the particle is hidden to some extent by the wave, or the wave is an extension of the particle (and vice versa), we cannot predict where the particle will be at any particular time. This puzzle of location differentiates quantum mechanics from classical physics, which holds that a particle cannot be in two places at any one time. In quantum mechanics, all is a ferment of potentialities, possibilities and probabilities.

In 1927 Werner Heisenberg proposed that there were discernible limits to measuring the properties of fundamental particles exactly. According to his principle of indeterminacy or uncertainty, measuring the characteristics of a quantum object such as an electron always involves an element of uncertainty. The act of measuring itself has a disturbing influence. To see where a particle is, one must shine a light on it, thus striking it with energetic photons, and the particle's position is revealed by the light reflected from its surface. But each photon of light possesses momentum, some of which is transferred to the particle that is being observed. It is impossible, therefore, to determine the position of this particle without giving it an uncertain amount of momentum. We cannot measure this momentum without increasing the uncertainty of the particle's position. Thus, for example, it is not possible to determine both the exact position and the precise velocity of an electron at one time. The more precisely an object's position is measured, the less precisely its velocity can be measured – and vice versa. Planck's quantum of action specifies the tiny amount of uncertainty involved. It is so small that we do not notice it when measuring objects of any everyday magnitude, but the uncertainty principle applies to the measurement of all objects. This is analogous to the superposition of will in our state of being whereby we form our own reality.

This reality manifests within the framework of pre-set and immutable cosmic laws, which limit our exercise of absolute freedom and yet afford some personal freedom for making decisions. Human freedom, operating through our unique ability to decide freely (the human free will that is unlike an animal's subjugation to inclinations and predispositions), works within the pre-established framework, and we cannot act upon the cosmic structure which enacts the laws. For example, the wisdom that governs the cosmic laws, laws that harmonize with and complement each other in such a perfect manner, guaranteeing each person and each object his or its space and so on, is way beyond the capabilities that a human being can deploy. Furthermore, we cannot alter the reality to be able to reap rye where we sow wheat. Each species or entity is only able to embrace a reality appropriate to its nature. Thought, volition and action issuing from each person act on a cosmic stream of energy and give rise to ethereal forms through the activity of the cosmic laws. These finer material forms may eventually condense into manifest physical reality in the world around us. If we describe our reality or a shared experience to another person, using our common language, they will instantly recognize it but only according to *their* reality, not ours. Colour, shape and tone always remain the same, but the way they are experienced will be different for each person.

Indeterminacy occurs with other complementary pairings, for example time and energy. A relationship of uncertainty exists between accurately measuring energy and the time it takes to make the measurement. Fix one variable and you free another – and vice versa. The briefer the process of measuring the energy of an electron, the more limited the accuracy of the measurement. This implies that energy held by a particle can fluctuate within a short period so that a quick 'snapshot' measurement can pick up any value within a widely fluctuating range. If time and energy relations are uncertain, what impact does this have on operations at the subatomic level?

One answer to this may be found through quantum mechanics. The classic concept of conservation of energy still holds at the end of the day, but due to the inherent 'fuzziness' of the quantum level, the uncertainty principle permits that over very short periods and short distances energy can be 'borrowed' to overcome otherwise insurmountable barriers. The shorter the time over which this occurs, the more the energy that can be 'borrowed'. This means that by trading off less uncertainty in some parts of a system, it can make use of greater uncertainty in other parts. It is thus possible for an electron to 'borrow' and 'pay back' some extra energy within a small time limit whereby one value increases at the expense of the other. It is possible for a particle to 'borrow' energy against time and, according to the Planck quantum of action, the more energy that is 'borrowed', the shorter the time in which it *is* 'repaid', and vice versa.

Ordinarily, when you view one of the sides of a fish tank from the front, it will look like a mirror, reflecting the back of the tank, with nothing showing through the side. The light is bent as it passes through the glass and water because it can only travel slowly through them. In water, it travels so slowly that it bends significantly, thus it is reflected off the glass at the back. If you place a thumb against the side of the tank, then, unlike a real mirror, you may see the thumb through the glass as a ghostly image. The light in this area is being absorbed rather than fully reflected. This only happens if the glass is thin enough for the light to pass through it quickly, because it can then 'borrow' extra energy, absorb it, and 'repay' it in a sufficiently short time period. The thinner the glass, the more quickly light can pass through it and the shorter the time available for repaying the 'borrowed' energy. With thicker glass, the light takes longer to pass through, leaving too much time to repay any 'borrowed' energy, therefore it is not borrowed at all.

In applying indeterminacy to space and energy, the smaller the observed space, the greater the energy credit obtained. With this extra energy it is possible for particles to jump seemingly impossible energy hurdles that block or restrict their movement. Inside the atomic nucleus positively charged protons are held together by the strong force against electrical repulsion from their similar charge. This strong nuclear force prevails at extremely short distances, beyond which electromagnetic force would push the protons apart. Protons can be imagined rolling about in a pot, having too little energy to roll out.

The ways in which quantum theory attempts to measure the seemingly immeasurable, to pin matter down and to distinguish substance from process, can perhaps best be understood by analogy. How does a particle of energy escape from matter, for example during nuclear emissions in radioactive decay? The answer is by tunnelling, like a mole, through the potential energy barrier generated by the atom's nuclear forces. Because of its wave-like properties, every particle has its location distributed over a range of probabilities, rather than being characterized by a precise location. This probability may embrace the 'other' side of a 'barrier'. Thus outside the nucleus, the particle possesses seven times less energy than the energy hurdle in the nucleus. Protons in the core of the sun are able to penetrate the otherwise impenetrable energy barrier and fuse into helium. Fusion energy radiates as sunlight from the solar surface.

The idea of quantum tunnelling, stemming from Heisenberg's uncertainty principle, whereby matter can move from one spot to another without going through the intervening space, has been applied to technology with profound effects. For example, it has led to the development of scanners with a power of magnification of up to one hundred million. These can be used to measure even the weakest magnetic fields such as those produced by the brain (commonly referred to as brain currents), and have enabled us to understand how the sun's heat is produced, and what happens in the process of radioactive decay.

Vanishing straits

Did the universe emerge from a vacuum or from a primordial seed? Both are metaphors provided by the intellect straining to seize the intangible. We may well ask, then, why humans should wish to, or, rather, be *impelled* to attempt to grasp, with eye, hand or brain, that which is in essence beyond material time and space.

The intellect is earth-bound, socially and culturally conditioned in its natural limitation to understanding earthly conceptions of time and space. It is unable to fathom anything beyond these boundaries. It is incapable of intuitive perception, which, in pure form, is uninfluenced and immune from social or cultural conditioning, and which provides a wider experience beyond matter. Intuitive perception issues from our living core of spirit and is the *actual* expression of our autonomy of free will. Beautiful and brilliant as our brain might be, it is not equipped to enlighten us on eternal truths. Whether as an artist, philosopher, mathematician, scientist, sage or spiritual leader, man through the ages has perceived a more perfect world, of which this physical world is but a coarse facsimile. Whether it is labelled Heaven, Paradise, Elysium, Valhalla or some other name, it is the same. From simple beginnings there evolve, over incomprehensible time scales, semblances of ideal prototypes that are the living reality above the visible, tangible world – ourselves included.

From time to time we all experience impulses of what is seemingly rooted beyond physical phenomena. At our most aware, we have intimations of our immortality. But for the most part we find it easier to ignore these supra-earthly visitings, impulses from we know not where, things which we feel – wrongly – that we can do nothing about and that do not greatly impinge on our waking everyday lives.

The expression 'gut feeling' is no mere metaphor. Prototypes of the physical body come into being in a finer, astral plane of less dense material consistency. Thoughts take on form in the same consistency, energized by the neutral power that pervades the cosmos. We bear an inner body of astral substance, attached by a cord of radiation to the physical body at the solar plexus, a network of intercommunicating autonomic nerve fibres in front of the upper abdominal aorta. Here our living inner core of spirit is able to connect with the physical body – this is our gut feeling, or, more precisely, our intuition. This cord of radiation is often referred to as the 'silver cord' as described by some clairvoyants. The final, irreversible separation of the spirit together with its finer coverings from the physical body (in death) occurs when this cord is severed. As long as this cord remains intact, and does so for a number of days after the declaration of death, the spirit may still be sensitive to what happens to the body, even if the brainstem has ceased functioning (a medical criterion for death). People contemplating organ donation or cremation should take this into consideration.

The word 'soul' has been in use down the ages. It has indeed been found to be essential to most far-reaching studies of human behaviour. But there is still a lot of confusion as to what the *soul* actually is. For some, the soul is the actual core of man, his innermost essence, while spirit is often associated with intellectual activity, or perhaps labelled the unconscious mind. Others believe the soul and spirit dwell side by side. Actually, as explained in *In the Light of Truth: The Grail Message*, the soul is to be considered as the spirit enveloped by its coverings or cloaks except for the physical body. The soul is the form in which the human being appears before and after his earthly existence.

We share the physical body with animals, but not our living core, which is spirit. Pleasure – and the sense of well-being – are instincts that need to be gratified; they are in their way both desirable and sufficient for animal nature. Setting one's existence by them, however, can be degrading to our sublime core of spirit. Instinct describes unreasoned behaviour patterns that arise from the body; intuition is recognition made by the higher spirit, with insights that erupt unbidden in conscious awareness to resolve some issue or to channel creativity.

FOUR

Spirit

Any action, unless it is pure reflex, is preceded by decision. Naturally we have impulses where strong feelings such as grief, anger or joy will make mental shortcuts to the decision-making process, where we may shed tears or clench our fists, but for the most part we choose to follow a certain course of action. As Hamlet said, every man has business or desires, such as they are. We have conscious and unconscious needs, which bring about the simplest reflex to the most creative acts. We have to satisfy our feelings of duty to our family, our profession, our self-image, our King, Queen and country, our ideals and our sense of destiny. We all, in some degree, have people to protect, and ideals, values and principles to defend. We all have needs. But legitimate needs can and frequently do result in reprehensible deeds.

But how do we know what is 'reprehensible'? In a dog-eat-dog world, with competition for space and life-giving resources, where one person's freedom begins where another's ends, or ends where external realities intervene, where one's needs or ideals clash with those of one's neighbour, where the struggle for existence means that some must be eliminated, does the term 'reprehensible deed' make any sense? The answer is: Yes, it does – if only by virtue of the fact that we have found it necessary to coin the phrase in the first place. How do we know that eating, or torturing or killing people is wrong? *Because* we ask the question. Some awareness deep within tells us that certain acts are right or wrong. Of course, despite such moral imperative or impulse, one can perceive the rightness of taking action and either do nothing in a moral fog or courageously set about to do what is right accordingly.

Moral law is a human concept. There are natural laws, which exist independently of our thought and action, and which are the external realities with which we interact in shaping experience. There are also human laws that we derive from instinct and from intuitive sensing in shaping society and culture. Natural law, demonstrated by science in countless mosaics of sequenced patterns, presupposes there is a Law Giver. Or else – no law, no cosmos. Everything in the physical world is interconnected. The greatest is mirrored in the simplest. The most massive and distant star is subject to the same law as the closest and tiniest particle. The natural laws interact to very fine and precise coordinates to generate galaxies, suns and planets, and enable plants, animals and humans to come into existence. Human law, expressing our inner regard for others, is no less guided and far-reaching. Rudiments of moral instinct are evident in other creatures, in dolphins and whales, for example, who play and grieve and show mutual respect for each other, much like ourselves. All of which might suggest that both human and natural laws originate from the same source, that they are subject to universal principle and will. Of course, because we humans have free will, we can make laws that are inherently unjust, but this can only happen if the basis of our laws becomes self-interest or simple error, or the pursuit of some 'ideology' or the struggle for power, all under the motto that 'the ends justify the means', that is, the domination of the intellect to the total exclusion of any intuitive feeling. Thus it was quite legal to burn witches and heretics in medieval Europe, to buy slaves in pre-civil war America, and to gas Jewish people in Nazi Germany, just as it is legal today to stone people to death or to severely beat or mutilate them in some Islamic countries for what sometimes appear to others as minor offences, or to keep terrorism suspects imprisoned for years on end in remote locations in wire cages without charge or trial. These 'laws' justly cause outrage when their morality is scrutinized, and they are very often viewed as criminal acts whatever their legal basis.

Tentacled eye

The concept of a law implies the existence of a Law Giver, as stated before. Rejection of the roadmap of faith for rational 'truths' of science steers thinking in the direction of atheism, even if it is on the basis of a secular search for personal development and self-realization. In removing God from the way things work in the world, science has left us only a series of baffling questions, which, in turn, have become our life's work.

These questions, once sinful to think about, let alone to ask about, are now open for debate. Does God really concern Himself with every single one of us, weighing our thoughts and deeds to reward and to punish? Are our experiences any greater with all the thrills and chills of charismatic religion? Does God intervene at all in our personal fate? The responsibility is on us to find the answers to these questions.

Why would God allow all the misery, sickness, injustices, crimes, and suffering on earth? What is the origin of evil? Why did God not create perfect human beings at the very outset? Why does He not intervene with order and happiness for all? How are we to understand His much-praised Omnipotence, Love and Justice, or His Omnipresence and Omniscience? Does God even exist?

Today belief matters to fewer and fewer people. Many communities believe we can get by quite well without God. The concept of God is for some no more than a 'figure of speech' for expressing anger, indignation, surprise or relief, thus invoking the 'forbidden' or 'taboo' – a primitive necessity that is still present in our psyche. The more our concept of God fades, the more we misuse or abuse His name. There is, of course, a terrible irony to this, a visible sign of our decline. Even films classified as suitable for viewing by children contain these 'figures of speech'. Is it any wonder then, that toddlers who are barely able to string three words together, become 'fluent' in these swearing utterances? We seem to both love and need taboo words.

Stirring moments of greater awareness, when we are especially close to the deep questions and the meaning of life, are swept aside in our daily round, never to be mentioned in polite conversation, discussion of them having been made taboo. About such things, we cannot know (and should not care) – these seem to be the maxims underlying most of our busy lives. Caught up in the business of getting and spending, the majority of people shy away from dwelling on thoughts of approaching death, and the hereafter and the central question of what life may possibly expect from us or what our purpose might be.

Unlike the beasts, we are aware of our mortality. A child eventually comes to realize that death is certain. As an adult he will adopt strategies for distancing himself from this knowledge. In his declining years, when intimations of death's ever-present proximity press with ever more precision, we either anticipate release from life's burdens or we dream of personal survival hereafter. We have our methods of survival, growing into that character that allows us to act defiantly and to survive our frailties.

A serious-thinking person can hardly relate to the experience of God with the kind of idealized human characteristics and superhuman powers that theologians invoke: an all-forgiving Deity who is privy to our thought and our actions, an Almighty Avenger who is able to punish and reward, trusting His representatives on earth, the clerical dignitaries, unconditionally to fulfil His Will.

The world remains at an impasse, in as much as the sense of an ending has always dogged the way humans think about the world. A certain span of time is pre-ordained for our development on earth. Today, however, events speak with unusual disquiet and pressure in the demands of the time. The Last Judgement presages the denouement to our passage through the material spheres. It is analogous to the final examination at the end of our schooling. For many people, then, science has been turned into the sovereign quest for all knowledge. Its enterprise between the senses and the mind, between the world of appearances and the world of ideas, has become, one might almost say, the new religion. It may even offer a more certain access to God than religion itself. Indeed, this seems to be so, as many of our greatest scientists – from Kepler to Newton, from Planck to Einstein – all had a profound belief in God. It was obvious to some of these that science, unaided beyond resources of the brain, can probably do no better than provide a tool to clarify and sort perceptions generated intuitively, and so long as it remains exclusively bound to the intellect it must prove defective in seeking to assume authority of knowledge. It seems that only the lesser scientists, the ones who cannot see the bigger picture, the ones who 'cannot see the wood for the trees', as the saying goes, are gyrating towards atheism.

Traditionally, the premise of God has been appropriated to religion, as if religion were a hermetically sealed compartment from which the air of science, or even the light of common sense, must be excluded. This mystical box of religion is kept for special purposes and reached for only in times of dire necessity, when the limitations of our human powers are exposed, as, for instance in times of life-threatening illness, or when reeling from natural disasters or on the death of a loved one. On these stressful occasions we are snatched out of the daily round and confronted with a void. We seesaw between hope and despair, often helplessly giving way to uninhibited pleading. Self-pity, introspection and retrospection apart, for the most part we find nothing that is not already within ourselves, no answer from 'out there' or 'up there', nothing to give comfort 'from above'.

Belief in the power of a God or some all-pervading presence might well become lost, for the moment, or for all time. The only truths might seem to be the reasoned abstractions of science in prying out nature's secrets, but these offer no creative basis for resolving questions of action and choice. The only reality might seem to be that of matter, in terms of measurable forces that are accessible to our technological control, and the material realm might then seem to be no longer an arena of divine transcendent ordinance. What is more, the world of which we are a part might be made up of aimlessly drifting atoms, some of which, by blind chance, might aggregate into 'life' forms.

But why must it be 'either/or'? Why not 'both/and'? For surely scientific knowledge and religious knowledge are *complementary*, not antagonistic in apprehending the world. They are convergent aspects of the truth, each needing the other, neither complete in itself. The recognition of the one Truth that is the firm guidance for every happening, every action and choice, has not been achieved by the religious traditions, let alone by theology and science, regardless of whether they are working in unison or opposing each other. On the other hand, the simplest of minds can feel the truth as revealed by seer, prophet or poet, or even by any other simple honest-to-God man. For while knowledge (the domain of the intellect) is acquired, understanding is revealed, and this process of revelation can only be intuitively felt, as it is the expressive domain of the living spirit.

Today we tend to put much emphasis on the need to acquire new skills, with the latest technology and the fastest internet connections. We emphasize speed and ingenuity and are obsessed with numbers, forgetting that these are mere symbols. What about old skills? Old ways of knowing? They are not necessarily decaying; they may have something to teach us. What is the value of speed if you do not know where you are heading?

Thinking makes it so

There is no objective proof or disproof of the existence of God. There are plausible arguments and many hints and testimonies, and evidence such as the shroud that covered the body of Jesus. Much of this evidence is mere oral testimony, some hysterical or manifestly fake. We can always find rational evidence for our arguments, whether the bones of Piltdown Man or the *Dead Sea Scrolls*.

If we direct our consciousness towards something, we give birth to it. In every moment of our waking life all our reality is pulsating with motion. We are permeated by the breath of a mighty life-force, which we may steer creatively toward a certain expressive form, according to the autonomy and nature of our volition. This ubiquitous spiritual energy stream does not originate from us, but rather it is at the disposal of our volition, thus our action and choice toward a goal or a value. This implies, of course, that we are answerable to the owner in terms of what use we make of it, thus we are spiritually responsible for our volition, and there must be an owner or source, because to assume that something can be dispensed from nothing defies even our own best logic.

Thus we are possessed by this neutral power of life which streams through all there is. We use its inexhaustible energies every day without giving it a single thought, but we could do *nothing* without it. Yet this force is incomprehensible to us. What, we may wonder, prevents us from calling the all-embracing life-force the power of God? Or if we believe that the entire Creation was formed out of itself by a random Big Bang event from a vacuum (without some will to give it direction or to light the Big Bang fireball), what was there before?

Who or what created the prerequisites for the Big Bang to occur? Could there perhaps be a different consistency or dimension of space and time before the Big Bang, the universe blazing into existence in a chain reaction from quantum fluctuations radiating out of 'nowhere'? These are of course unsolved age-old questions, loaded these days with modern knowledge. In a world that was truly godless, and thus however apparently comprehensible it might be it is still devoid of direction and purpose beyond a human construct, we would not be wrestling with these problems. We would not be agonizing, spending time talking about the Will of God.

Our physical senses perceive material things, tying us to what the brain registers in a world bound by space and time. But we know there is finer reality beyond the world of appearances, the nature of which is beyond perception of the physical senses, beyond the grasp of the intellect. Why this should be, we do not know. We may try to label it, as is our human custom, invoking by it the Great Spirit or the Will of God. The name without understanding is meaningless. We bear within us a free-floating spark of spirit that in its intuitive perception transcends the rational boundaries of space and time, one obedient to laws other than those that are consistent with the purely physical.

Freethinkers prefer to speak of the 'law of nature' or the 'life force', rather than the Will of God. The disparity of terms does not necessarily indicate two fundamentally incompatible ways of looking at the world. There are, as we have said, no empirical proofs of the Creator's existence, yet we are surrounded by endless marvels, not least of which are ourselves and the most finely tuned forces that undergird the universe. By *proof* we mean that which our rational minds can grasp. We can no more resolve what is beyond the scope of our senses in this way, than a sieve can scoop water. As Walt Whitman says, a mouse is miracle enough to stagger sextillions of infidels. And that is no mere flight of fancy.

Today we are sceptical of miracles, we regard them as freak events that one cannot explain in cause-and-effect terms, and declare that 'miracles' these days are becoming fewer. Yet fresh miracles appear on the horizon every morning. The big question remains, to which we have found no answer: that is, when, where and how could such things happen without will or purpose?

Ideas about God come in many forms; He is seen by turns as all-knowing, all-loving, ever-just, or always concerned, yet also vengeful, jealous, terrifying and even merciless to the disobedient. Like the Greek deities, the biblical God seems much like us except that He is more powerful, one who in the manner of the pagan gods must be propitiated with sacraments, penance, sacrifices, obeisance and prayer. Plainly this is anthropomorphism of the highest degree. These notions, though understandable as human projections, exemplify very primitive thinking, which cannot be reconciled with the new insights into the world gained by science. Thus many people retreat from being convinced about the existence of God simply because the caricatures of the Creator presented to them by the ecclesiastic authority lack internal consistency. Unfortunately, as people are apt to do, they overstep the mark and 'throw the baby out with the bathwater'.

The traditional God is all things to all men. His justice is administered, we hope, not too sternly, for that would stand in the way of love. His wisdom ranges from knowledge of our every thought to knowing the exact number of stars twinkling in the great canopy of the firmament. Basically we insinuate what are thoroughly human characteristics and cling to the idea of a benevolent God within reach, one who will help us when we summon Him, who is hankering for our acceptance and is represented by elected human beings.

The culmination of all that is outrageous in the distorted conceptions of God lies in the sublimation of how God works to feudal authority. Ecclesiastic and secular authorities have laid down in His name what is a sin and what passes for moral law. These authorities, being authorized by God, are – so they tell us – empowered to forgive sins and underwrite state punishments. They prescribe what we should believe and make it a sin even to reflect on their doctrines, or even subject them to keen scrutiny. Even after the Reformation, when religious freedom was finally won across much of Europe, this applied only to the ruling classes. The common folk were forced to adopt the religion of their feudal masters. Failure to do so resulted in banishment.

To attribute to God such narrow human concerns is to debase ethics and morality. The perpetration of such mummified falsities is enough to encourage the illusion that one can manage without God, or is it merely an intuitive rejection of these falsehoods taken too far, as stated previously?

Belief in monotheism, from its Judaic roots to Christianity and Islam, has frequently been impressed on subjects with brutal force and excessive vigour. The cliché of the *Bible* in one hand and the gun in the other lives long in the memory and still resounds today. Despite this, a belief in an impalpable Being and a Primordial Power or Light remains a source of continued vitality and residual strength in human consciousness. The conception of 'God', a 'Supreme Being' or 'Great Spirit' lives on despite human distortions and a steady decline in religious observance. Is it that we cannot avoid – or are programmed by our genes to adopt – anthropomorphic notions of a Creator?

Conceptions of God have changed radically over the centuries. The vengeful, quarrelsome, wrathful Jehovah of the *Old Testament* that emerged in the face of human state of inherited sinfulness and flawed ethical pedigree was succeeded by the moral authority of love and justice in the Christian order. Today, a new outlook exists beyond the fragmentation of denominational ideologies. Just as Auguste Comte, Thomas Carlyle and many others had to throw off those old Hebrew clothes, and George Eliot freed herself from that 'wretched Giant's bed of dogma', so modern man has to change clothes, to put on the shining white suit of scientific thought.

Today it is, perhaps quixotically, to the scientist rather than the cleric that we may look for enlightenment about the Divine Will of God. The nineteenth-century battle between science and religion has been fought and won. If by the mid-twentieth century science appeared to be victorious on all fronts, its triumph now seems a Pyrrhic victory. In war there are no winners, only losers. The geneticists and cosmologists of today have found new worlds, new ways of seeing, of demonstrating to us miracles of life, space and time that seem even more staggering and exalted than the mystical raptures trotted out by the church to dazzle the populace. This is not a case of God 'making a come-back', but rather of science opening our eyes a little more to what has always been there, as it were staring us in the face. Modern science is getting ever more transcendent, the observations ever more surreal and paradoxical as physics and metaphysics touch one another. There is no surprise in this, because there is no division between this world and the beyond. Only the veils of the senses, or the limitations of measurement, appear to separate the spheres. The rational evidence points the way to the immaterial Will that manifests in the immutable cosmic laws.

Cosmologists will not come up with the medieval figure of an arbitrary God. They will meet the working of an Almighty Creator, whose perfect Will rests in Creation as the omnipotent and unchangeable laws of nature and from which, for example, Copernicus, Galileo, Newton, and others have lifted a veil. These active laws are expressed uniformly in the entire Creation according to the consistency of each sphere or plane. They are the laws of development that are completely responsible for the processes of this world and at the same time act automatically with immutable justice to return to the person in abundance the fruits of his or her deeds.

Sounding the depths of time

The word 'law' has various meanings, from the principles behind mathematical formulae to the concepts and categories of jurisprudence. The Law of the Will of the Creator is not restricted to established or limited forms, but enfolds in abilities, becoming manifest in material and spiritual world events. Science cannot exist without laws and we could not come to understand the spiritual without knowing the laws that govern it.

We have the capacity to recognize the Will of God. Emissaries have incarnated at all times among various cultures and communities in the course of human development as helpers towards recognition of the Truth. Three basic laws gear into one another, harmonize and complement each other in the perfect mechanism of the entire Creation: These are the Law of Attraction of Homogeneous Species, the Law of Reciprocal Action, and the Law of Gravity.

The Law of Attraction of Homogeneous Species is exemplified in the saying: 'Birds of a feather flock together' or 'The apple falls not far from the tree'. We see it at work with people who are 'on the same wavelength'. In large and small human groupings, in families, like-minded people assemble. Societies form, with castes, social classes, political parties and interest groups. Smokers come together, drinkers find one another, gamblers associate with gamblers, and so on. Idealists are linked, and so too are anarchists. Homogeneous thoughts unite in strength, and in this way we can be influenced by unconscious links with other thoughts that affect our sensitivities, attitudes, behaviour and actions.

Thought forms influence the will-power of people who are similarly disposed. Our volition moulds the neutral cosmic power currents into living forms that attract what is homogeneous. Hatred, envy, intolerance, oppression and the like are forms that are most prevalent, including the demons and evil spirits known to primitive cultures. Even in our 'enlightened' cultures, such things are hinted at in the saying 'confronting one's demons', whereby one's own thought creations can gain tremendous power by attracting their similar kinds. Unfortunately our spiritual ambivalence gives rise to only weak homogeneities of love, kindness and compassion. Such dispositions are not as strongly expressed and not prevalent enough to act as a strong counterweight to the evil thought formations. Because of this imbalance and the subtlety of unnoticed influences, every one of us is constantly bombarded with profusions of evil thought forms, which try to attach themselves to us. People who are weak-willed or whose psyche is largely beholden to their passions, but who may not necessarily be bad, will process information from the assailing influence. What is more, outside their awareness, they will eventually succumb to such an influence. Occasionally someone behaves out of character and commits an act of violence, linked and in part driven by the energy of homogeneous thoughts, without awareness. There are some very revealing after-the-fact explanations in such cases, seen through statements such as: 'I was no longer myself!', 'I don't know what came over me!', 'I can't explain what took possession of me and made me do it', and 'Something or the other got hold of me – it was stronger than me'. Protective freedom from these attacks comes from a strong and unwaning volition for what is good within the context of spiritual vigilance.

Let us look at the Law of the Attraction of Homogeneous Species. This law manifests in the union of radiations in the atom, the union of atoms and molecules in crystals and in the growth of natural solids, as well as the formation of cosmic bodies on scales of daunting complexity. The law underscores attraction of opposite magnetic poles, and the urge for complementary union of male and female human spirits, whereby the split species of our male and female genders seek union into a whole. The law operates on whole rather than split species, whereby a male human being and a female human being are urged to unite to form a complete whole. This does not necessitate marriage or close contact, but shows itself in the type of activity of each respective gender, which must be in harmony for a society to thrive. Today we see considerable overlap in the spheres of male and female activity, which is an impediment to the necessary harmony. In many ways men and women have become rivals, especially in the workplace and in public life and, for as long as they both want to dominate, they see each other as threats rather than complementary partners in a broader sense that each cannot do without. Struggle for life and for dominance ends up working against cooperative urges.

Our urge to seek out those who are like-minded and complementary is an indication that no person is complete on his or her own and that we need other human beings, particularly those who can help us to outreach our earthly bonds. Thus the law of homogeneity works in two ways: firstly, it causes like-minded people to come together to further their goals; secondly, as a derivative of the primary law, it causes people to seek out others of the opposite sex with the same goals but with characteristics that are complementary to their own for a more intimate union. Survival of our species also depends on this.

These are elective affinities, then, but for strength and development, differences may also be present – in sex, age, temperament, and even social class. For homogeneity must always tend towards the inclusive and shun the exclusive.

Every society divides into social classes, even if only in a subtle way and even if not admitted openly. The words 'class' and 'class struggle' are now loaded with baggage from former times and are therefore mostly avoided. On top of this, everybody wants to be seen as 'middle class'. Nevertheless, for a healthy society, the classes, whether their existence is acknowledged or not, must work together with mutual respect, each held worthy of regard. Each class, while avoiding feelings of superiority or inferiority, contributes its own valued artefacts and achievements, values not possessed by others but needed for a wholesome community. A healthy society depends, therefore, on complementary diversity.

The clustering of familial affinities, such as musical, artistic or mathematical talents, arises from the attraction of similarly striving spirits, who, in coming together, can further each other. Base inclinations are clustered in the same way. Circumstances of birth are also influenced by connecting threads of fate through another law – the Law of Reciprocal Action. By the time of our birth, we are already entangled in a weave of intricate relationships of blood that paves our immediate path in the earth-life.

The second law at play in the mechanism of Creation is the *Law of Reciprocal Action.* This law brings the harvest of deeds. To Christians it is the law of sowing and reaping, to Buddhists it is the law of karma, to scientists it is the law of cause and effect that will, in turn, generate a new cause. The principle operates uniformly in all Creation, in the physical and in the finer worlds up to the spiritual planes. It applies to the seed we plant in the earth as to our volition, thoughts and actions, which issue from us as seeds of a finer nature. Every deed that emanates from us takes a form of one kind or the other on finer or less-fine planes in accordance with its nature. The unavoidable harvest comes back to us greatly multiplied, because on its way the deed is strengthened by the attraction of its similar kind. Good thoughts and deeds will yield an abundant harvest of good; evil volition burdens with guilt and returns a harvest of evil. We cannot harvest rye where we sow wheat any more than we can reap goodness from evil deeds. This Law resounds in the deep truths of many idioms: 'What you sow, you shall also reap', 'What goes around comes around', 'As the seed, so is the harvest', 'Justice comes full circle', 'Do as you would be done by' and 'As a man calls into the woods, so it echoes back'. We are responsible for everything we do; responsibility sets in with the volition. We stand always in the loop of the threads of our deeds and their reciprocal effects, and send our works with us after death into the beyond, where we will be confronted by them, even if we had been able to evade earthly responsibility. Someone who recognizes his guilt is open to atonement by way of reciprocal action, which forms part of so-called 'fate'. His inward change for the good is able to lighten his burden of fate, which can otherwise further entangle him, and a serious inner commitment to good can help redeem the burden of guilt before the cycle of reciprocal action closes in. His lighter inner disposition will then weaken the returning blows of fate, because they no longer meet a soil of homogeneity within him, and, in balance, he is left with a symbolic redemption, of which there are many forms. Thus a small act of service voluntarily accorded to a fellow human being out of love or compassion may redeem a mighty karma or reciprocal action that would otherwise have hit the guilty person with full force and caused him much grief.

From physics and chemistry we learn that forces of nature unite the physical world. Whether it is a force that sets off chemical processes or governs coherence of the atomic nucleus or electromagnetic action between charged particles, there is no movement without reciprocal action. Every cause has a corresponding effect, which in turn drives a new cause, and so on – unceasingly.

When something happens we look for a cause, and expect the connection between the cause and its effect to be at least open to discovery. Every action has a reciprocal effect, even if not always discerned. Effects may occur on a large scale, as with industrial pollution, or on a finer level with movements of atomic particles. On an even finer level, they may occur with our private thoughts and attitudes.

Thoughts precede our actions and choices. Thoughts are regarded as free and are often no more than a pastime because we can think about anything, creatively or destructively. However, thoughts lead to words, attitudes or actions that can affect the thoughts, attitudes and actions of others. In this way, our thoughts make us partly responsible for the actions of others, possibly those we would never commit ourselves, and which we might condemn in the strongest terms. In this way the acts of terror and genocide we so strongly condemn, incomprehensible to us in their brutality, are fuelled in part by the thoughts of ethnic, tribal, racial and religious bias sent out constantly in an unrelenting swarm by so many people.

In the Law of Reciprocal Action each deed will return to its originator, reinforced in multiple. We cannot reap kindness where we sow hatred, nor joy where we sow envy. The harvest arrives in abundance, even if the seed may have been sown in a former earth-life. How far is a single neo-Nazi fanatic responsible for the ideas implanted in him by preceding generations of racial supremacist thinkers? We are struck out of the blue by a 'stroke of good fortune' or a 'blow of fate' that we cannot explain and which we may ascribe to good or bad luck, respectively. The far-reaching importance of the law cannot be over-stated. Existence, in its multifaceted interconnections and interrelationships, is no arbitrary accumulation of actualities nor is it dependent upon contingencies; it follows immutable laws, which work to the freedom of everyone.

Giving heed to the law of reciprocal action opens us to becoming more content with our lot in life. An isolated person will be lonely. A dishonest person will perceive the world as a den of liars and cheats; 'What you sow, you shall also reap'. This binding moral necessity is not just a Christian message. It is an inherent part of the stream of life, manifest in a prime law of nature. Ignorance of this law may breed doubt and make some people think it is possible to reap benefits without sowing. We see the world as we are.

Whoever wants to be loved must himself express love. Whoever wants attention must be able to listen. Our fate is spawned by reciprocal action. Every thought, word or action emanating from us will return to us greatly multiplied in its effects. We can change our way of thinking, view the greater and lesser joys and tragedies of our life with the right prism of acceptance, atone for past deeds, and free ourselves from negative feelings or feelings of guilt. But no person can expiate another's guilt; each must make the change in conscience or attitude for himself. The idea of expiatory sacrifice by one person for another's guilt, exemplified in Christian doctrine, is a false doctrine. It takes away responsibility. Viewed without prejudice, the very idea is strange. What would we think of a judge who knowingly sentences an innocent person while letting the guilty one go free? And what is more, might we venture to expect this from God, who Himself *is* Justice?

Once an action or choice is set in motion the consequences are beyond our control, often unseen and unforeseeable. Like a stone that is thrown and, unknown to the thrower, is heading towards a pane of glass. But the decision to act precedes the action. It is here that we can exercise free will; once the action or choice is made, the irrevocable lawful process of reciprocal action is inescapable. To illustrate this point, you may decide to sow corn rather than any other grain, and once it is sown you cannot go against the reality manifest in the cosmic laws that make it impossible to reap oats. The autonomy of free will is inextricably bound up with personal responsibility. In this way – and only in this way, despite the working of the law – people can be seen as the 'architects of their own fortunes'. Free will can break the cycle of fate, negatively as well as positively, for example in symbolic redemption. To spell it out again: only the *decision* is free. The consequences arising from that decision will always follow the Law of Reciprocal Action and fall back on the originator, sooner or later, without fail. That is not a matter of hope – but a law!

Our ignorance of the workings of the laws of nature or the cosmic laws is partly due to the dominance of our intellect and erosion of the power of our critical intuitive faculty. The intellect is the perceptive capacity of the brain and is thus bound to earthly concepts of time and space. It is confined to perceptions of the physical world. In contrast, the intuitive capacity is capable of recognizing hidden transcendent connections with our being, a swarm of unseen causes and processes attaching to us. We often no longer comprehend concepts that express spiritual or divine working; although these ideas are naturally invisible, they exist nevertheless. An entity can only conform to its nature, and the intellect must conform to its material nature. Only earthly things can be ascertained by material means. People schooled exclusively in the intellect therefore dismiss spiritual activity (which is finer) as fantasy or superstition. Yet too much arid science is not good for the spirit – that is why we joke, write novels, and so on: to feed the spirit. Materialists deny the existence of a soul substance that is independent of the physical body, as well as the possibility of consciousness outside the confines of the physical senses and material reality. Modern thinking is over-dominated by the intellect and does not benefit the spiritual part of human beings. This narrows our comprehension and diminishes our self-awareness, both of which are certain obstacles to human progress.

The concept of reincarnation, once fashionable in a literal sense, but now seen more as allegory, corresponds to none other than the application of the laws we have been considering. The outlook of a single earth-life obscures one's grasp of the connections that weave our individual fates. There is a degree of truth in allegory, based on the supposition that happenings in the physical world are a replica of finer reality.

As below, so above. Thus we say 'history repeats itself' – with understandable modifications – and 'things come and go in cycles'. That is until the purpose is complete, often going beyond the brief span of one earth-life, and the circle is closed.

The third complementary law, the *Law of Gravity*, works in harmony with the Laws of Attraction of Homogeneous Species and Reciprocal Action. Gravity pervades the whole of Creation, stretching through all the finer and non-material planes to embrace the world we see. On earth, an object floats or sinks in water, and rises or falls in air, depending on its heaviness. Avalanches form from innocuous snowflakes. In the world unseen, the soul that has exited from the earth, thus detached from the physical body, is drawn by the finer process of this law of gravity to the plane of its homogeneity, where souls dwell with similar striving. Base propensities, in their heavy condensation, weigh down the ethereal body; the soul will accordingly sink to a homogeneously dense environment in the ethereal beyond. Striving for higher and purer goals will lighten the body. The spirit core itself exists in a natural state of weightlessness; density applies only to its coverings. A heavy ethereal body will result where the spirit has not exerted itself.

It is much more than a persistent metaphor, symbol or archetype that leads us to speak of the lightness or heaviness of spirit, which imbues feelings within us of being weighted down, sinking into the depths of depravity and depression, or rising to new heights of inspiration and floating on what we term as 'cloud nine'. Nobody will deny that sad, angry and depressive thoughts weigh strongly on our inner life, or that glad and cheerful thoughts are inwardly uplifting. It is true that nobody likes being depressed. Base propensities drag souls down to the homogeneous ethereal environment, where they live through mutual indulgence of the dense inclinations until, eventually, revulsion makes them yearn for something better. The nature of our ethereal environment is able to affect our psychological foundation and consciousness, even while we are on earth.

Imagine a woman who is moved by the words and music she hears at a concert. As spirit, she is lifted up during the performance from the lowlands of mundane striving and bustling to the point where she perhaps completely forgets herself. She leaves the concert hall in a state of animation, feeling her body in a totally different way, moved, as we rightly say, worlds away from her habitual body. She feels an enormous lightness, something like an out-of-body experience. This is the effect of the great principle of gravity. In general, a spirit-ruled body will make an almost weightless impression, with the finer and higher impulses holding sway over the lower, dense inclinations and predispositions, including passions.

Light gleams an instant

Matter is cosmic stream of energy, that has been cooled and condensed into a corresponding material form. Power streaming from the summit of Creation is slowed down and the ensuing condensation of energy gives rise to the material realms. The density of matter can be reduced by increasing the mobility of its constituent atoms and molecules, for example, by heating it up – thus, vapour or steam is less dense than water, and a metal like copper is less dense when warm, as it expands. Ice, although colder than water, is anomalously less dense than water. It floats on top of water as it forms because liquid water expands near freezing point; its molecules become more spaced out in its crystal lattice structure. This property means that fish and marine life exist (essential as they are to the food chain) because if the ice settled to the bottom, the overlying water would be exposed to the cold conditions and the whole body of water, whether stream, lake, river or ocean, would freeze up into a block of ice and any creatures within it would perish. In a hot-air balloon the air is rarefied by heating, and is capable of lifting people or equipment; cooling of the air leads to descent. Atomic fission releases enormous energy; thus we may say that matter is akin to a centre of 'frozen' or 'solidified' energy. Space and time, and the universe, are driven to expand by energy. What, then, is this abstract quantity called 'energy'?

In science, energy is seen in terms of the predictable activities of purely physical causes and effects. In reality, energy originates solely from spiritual power. When we come to speak of 'spirit' we tend to lose ourselves. While we recognize its presence everywhere, it is beyond the capacity of our intellects to grasp it. But it is important that we try to work with the indefinable, like the suggested forms in Leonardo da Vinci's line drawings of models, and be the stronger for it. Of necessity, our discussions about the spiritual workings of nature, although scientifically based, will reach beyond science. To some it may appear as metaphor, and to others a depiction of a higher reality and a higher law.

What we can be sure of is that spirit energy impels matter. However, there are gradations of spirit, just as there are gradations of matter. Thus the species of spirit that animates matter has a weaker power of attraction than the species that animates the human being. Spirit-particles of energy, streaming uninterruptedly from the summit of Creation, shower our cooled universe, which is characterized by dense matter and sluggish vibrancy.

Although human beings are outwardly of the same substance, inwardly we may be, as we say, worlds apart. Some souls are heavy or weighed down in spirit, some are dark and brooding, while others are light and less encumbered. The very limited changes that can be made to the physical body in a life-time can mask the more vibrant changes that may have taken place within. Inwardly we may be either Calibans or Ariels, outwardly beautiful and inwardly evil at the same time, or vice versa. Outer and inner affinities may be mismatched in a camouflage of the self. Moreover, depending on the demands of the situation, villains can act courteously while ordinary folks behave like villains, the unseen volition quite at odds with the expressed words and action. Many who have entered into a love relationship based on outward consideration, oblivious to (or minimizing) certain obvious or subtle, far-seeing clues to emotional discordance, may bear this out. Eventually a person's true nature shows because it cannot be hidden forever. Many disappointments would be avoided if we listened to our intuition's inner voice a little more in our actions and choices. The impressions formed within a few moments of meeting someone are usually right and lasting; this is when our intuition speaks to an immediate wider recognition of the nature of the other person, including aspects that are inaccessible to our intellects. Unfortunately, in many cases the intellectual sagacity of the other person may make us rationalize away the initial impression, but it often turns out, often to our great sorrow, that that first impression was correct after all. Once a soul shows, as it inevitably must, then incompatibility can no longer be disguised.

Having spoken of 'soul' one is almost obliged to consider those vexed places in religion to which the soul is inevitably bound – those places that are but states of being: Heaven and Hell. The soul may be striving upwards to the luminous origin of life or down away from it. It is the call of our free will. And these states are not entered into only with physical death; they permeate the whole of existence. However, we enter into the full state of our dispositions after discarding the shield of our physical bodies, together with our intellects, when we depart this life. We are drawn to and held fast by our band of homogeneity. As with gravity, sometimes we are 'heavy at heart', while at other times 'a stone falls from the heart'.

Nobody wishes to be weighed down or oppressed. Happiness consists of being light of heart, nearer to the source of life.

It is no accident that the word 'light' extends its meaning from weight to temperature; the warmer something is, the lighter it is, in both senses. Just as the light of the sun animates us, warming whatever it falls upon and freeing it from icy freeze, so there is also a finer light that animates and warms our soul. Like plants striving upward to the sunlight that sustains them, like the sunflower always faces the sun, so our souls strive towards spiritual light for invigoration.

In dreams begins responsibility

The cosmic laws explain the connections between our free will and our fate. We are free to sow, but are bound to the harvest. We are free to plant, but must savour the fruits. Free will is an ability of the spirit deep in us and is inextricably linked with responsibility. Without responsibility we could have no free will. If nothing or nobody could be held accountable for anything, chaos would reign, but the active cosmic laws, as the guardians of order, prevent this – of course – by giving each his due.

In truth, freedom is consonant only with the exercise of the free will of the spirit. Social provisions and guarantees of rights or freedom of belief, thought, expression, the right to assembly, and so on, merely serve to remove external obstacles to the individual's exercise of free will, but do not compel it. No-one is free if he does not exert the volition of his spirit, if he is not guided in action and choice by intuitive recognition, even if he lives in a society that guarantees a swathe of fundamental rights. He must also overcome any inner obstacle, some kind of propensity, to the free exercise of the volition of his spirit. A propensity might be a penchant for the lure of money, or envy, or hatred, and it will always overshadow a decision beforehand and steer it rigidly towards a particular direction, thus blocking and making the free exercise of his will a constant struggle.

The spirit, our core of being, should lead our decisions of will, while the brain should simply execute the volition. The spirit is a completely different species from the brain, and it is unable to interact directly with the earthly domain like the brain can because the brain – being a physical product – has a similar nature to the earthly domain. The brain works to express the choice of the spirit, which reaches consciousness in intuitive perception. And to accomplish this task, the brain likewise possesses the ability to decide, in this case through the intellect, so that it can make a choice among the multiple material options available.

But the more a person is subjugated to his intellect, the less he exercises the free will inherent in his spirit. And the less he uses his free will, the less he is free. No spirituality (not developing our intuitive abilities, for instance) means no freedom! In addition, the limited scope of reality means decisions that are made with the intellect and unguided by the spirit are liable to be fragmentary ones, often in some conflict with the cosmic laws. This again leads to a burden of fate.

Human laws on earth do not always harmonize with the cosmic laws. As stated previously, grave crimes have been sanctioned by earthly laws. But even if this is not the case and a law is fair and just, our judicial system can only assess the visible deed, the thought volition that preceded the deed being beyond its reach. In the spiritual sense, however, a person is already guilty because of the negative thought, even if circumstances prevent him or her from carrying out the deed. Such a thought may, through the principle of homogeneity, become entangled with and thereupon reinforce a complete stranger, who may then carry out the deed elsewhere.

All our deeds – intuitions, volition, thoughts, words and actions – are analogous to seeds that germinate on otherworldly planes and the fruits of which we must harvest, whether in this world or in the hereafter. But how we then react to our fate is left entirely to us. We can see injustice in fate and react negatively, perhaps thinking that we are destined to do things and are subject to circumstances completely beyond our control, or we see an opportunity to learn and to develop, and thus also determine what lies ahead for us. In any case, once we awaken within ourselves an earnest volition for what is good, the liberation from the shackles that we ourselves forged will surely follow.

The three cosmic laws described here maintain movement and development, and they also promote 'upbuilding' in the entire Creation. In addition, there is the Law of Equilibrium. This is related to the balance that is necessary between giving and taking, and it is also essential for harmony. All these laws unite in the Law of Motion. All of Creation is in motion.

Light forever shines

Across the universe, the greatest events are mirrored in the smallest. Electrons surround the nucleus in the atom at distances determined by their energy content. Science cannot tell what energy actually is made of or where it comes from, but we know that its animating action is connected with light. The constituent of light energy is the photon. The more photons an electron absorbs, the more energetic it becomes.

We can describe energy as a vibration that impels motion and change. Material interactions take many forms, but why or how energy works in the first instance is a mystery. Why did energy come into being at all – and when did it do so, in a timeless universe? Allegory begins here.

We begin and end with the question of light. The smallest components of matter become more real, more aglow with power, when they have absorbed light. Light is a life-affirming radiation of enormous vitality, vibrating billions of times per second. It is the expression of creative power within the biosphere. The example of the electron exemplifies the necessity to open the self to light in order to obtain energy. If our species is spiritual, as people have practically always felt in their heart of hearts, then 'light' in this context is also spiritual. We too are imbued with light. The conversion requires that we rightly know and understand that the mysterious energy that binds us with the Light, the origin of life, can only be spiritual. It is a stream of volition emanating from the summit of spiritual existence.

We see energy harnessed in the fallout from a carefully directed jet of water or in an atomic explosion, both involving precipitations of finest spray drizzle. What we call energy is the basis of structure and the forces between the smallest components of matter. We experience its effects in subatomic radioactivity and in colossal supernovae. It accounts for unceasing change and movement in the world and originates from spiritual power streaming from the summit of Creation, the archetype of all the planes that lie below. This may be cynically dismissed as a metaphor, but how else does one speak of the unspeakable, the power that moves all Creation?

There is a vital difference between the energy active in an atom compared to that in ourselves. The total energy of the electrons furiously 'buzzing' or spread around the nucleus determines the energy state of the atom. Electrons occupy levels based on their energy. As it is in small-scale matter, so it is also with us. Individuals of diverse experience co-exist on earth, bustling with an energy that may seem purposeless, or braced, as though, in suspended animation. However, there is a vital difference between the active energy in an atom compared to that in us. In atoms the spirit-core of the energy particle is without any volition and is no longer capable of developing beyond its present state, whereas our spiritual species has the capacity to mature to a self-conscious personality, thus the spirit's apprehension of itself.

If the different chemical elements along with their properties are compared to small groups of people, each group will be found to have its unique characteristics, and each will have a frequency distribution of those characteristics. Each atom will have one energy state, and atoms with low states will be in the majority. A transition in energy state requires absorption or emission of a quantum of energy. The more rare atoms, those that vibrate on a higher energy level, will remain elevated. Something similar occurs within groups of people, but with the crucial distinction that in their personal exercise of free will, individuals may exist only in their developed state of spiritual experience. So it is that the numbers of spiritually unenlightened people far outnumber those who are enlightened, largely because of the subjugation to the intellect and voluntary indolence of the spirit. Many people nowadays have an overcultivated intellect at the expense of their intuitive faculties, and their cognitive awareness is only a little ahead of the most highly developed animals. We are yet to absorb this lesson fully. We must try to understand that spiritual maturity has very little to do with a person's profession or career, or social standing. It does, however, have everything to do with attributes such as nobility of character, integrity, ethical behaviour, loyalty and compassion, to name just a few. A perceptive individual will find these people among all social classes because spiritual qualities, though not obtrusive, cannot be hidden.

The spiritual power of light, the nexus where matter and spirit cohere, is perhaps best illustrated by the phenomenon of lasers. A laser produces an intense light beam of a single colour. This laser beam is a product of the interaction between radiation and light. It is a union of an enormous quantity of photons in a single quantum or energy state. Large numbers of atoms with their electrons excited to high energy states are stimulated to emit photons all at once, the emitted waves being in phase and of a single frequency. This coherent beam can be focused into a tight stream that does not spread out, so it can be transmitted over great distances with very little scattering. The manipulation of atoms by laser beams has wide scientific, industrial, medical and commercial application, beginning with the humble CD or DVD player in a car, to optical computers and complex surgery conducted on the retina of the eye. Laser light can transmit undimmed impulses over interplanetary distances to satellite probes and cosmic bodies. It is not a natural occurrence, but the product of technology, constructed in many ways using various synthetic crystals.

The word 'laser' is an acronym for: light amplification by stimulated emission of radiation. The stimulation consists of over-turning the ratio of energy levels, so that most atoms of the crystal exist in a high-energy state. Some means are employed to excite the atoms by electromagnetic radiation into higher energy states. By exposing an ensemble of atoms to a beam of light of exactly the right frequency, each atom gives off a photon to the beam, in phase with the others. The bombarded atoms become revitalized, imbued with a new light and power, which is in turn reflected at the beam. Thus the original beam is amplified, so it becomes tighter and more powerful. However, to be effective the radiation must be of the right frequency, with the same vibration characteristics as the original material.

Applying the laser principle to people, allegorically, we may say that the different religious and cultural groups will absorb light only of their own kind and in their own way. And just the same as when an atom is bombarded with light of the correct frequency, some of its innermost electrons will absorb additional energy, so it is with the human will when it is bombarded with spiritual light.

Unlike the electron, however, which cannot retain its energy and must emit it again, the will of a human, when it is exerted, does not lose energy; in fact it gains energy and, like an electron bombarded by photons via laser light, it is able to galvanize others to a higher energy state. Benefit comes from not merely absorbing energy, but using it properly. It is indeed a duty to bring our inherent spiritual abilities to blossom to full realization from their original state. Not using the power means losing it. Thus we can imagine a sense of purpose and upward movement stimulated by light. Other stimuli, such as those from nicotine, drugs, alcohol and violent images, have no upbuilding purpose and, because they are self-directed, they are essentially self-destructive.

The stirring laser allegory points to the importance of our state of spiritual alertness and the direction which we give to our volition. Creation is permeated with light, with spiritual power, a power that is ethically neutral. We release the power through the volition of our spirit, tap into the power found in and beyond matter. That matter is influenced by the immaterial is shown, for example, by personal memory, recalled knowledge and skills, and by the influence of long-dead thinkers. Of course, these thinkers may be among us again in a new embodiment and the potentially humorous situation may arise from time to time whereby such a thinker may come to study or even criticize or repudiate his own works. This is why it is a great blessing that most people cannot recall their past lives, because if they did they could not approach matters such as these unencumbered and without prejudice. Our will can draw on the spiritual power for good or evil purposes. The choice is ours, but so is the responsibility that goes with it.

In a laser, light is added to atoms of a suitable substance, whose energy becomes raised and is stimulated and rectified. As a result, a light of great vitality, close to the highest material purity, is generated (or it might be more accurate to say released). This light manifests an abundance of qualities and effects. We would do well to heed the lesson of the laser, for it demonstrates the working of the great cosmic laws in bringing light to illuminate all things, including the enlightenment of the human mind.

Motion into time

Scientists suggest that there are immeasurable expanses of the universe, a real world out there that we can make sense of, in which according to the present state of knowledge there are more stars than all the grains of sand on all the beaches on earth. Theologians speak of 'heaven' in terms of a stellar expanse above this world. In the meantime, there are near-death, out-of-body experiences of the transition from this world to the hereafter to consider, which are well documented and systematically researched and therefore cannot simply be ignored or dismissed as fantasies generated by the brain. How does this all fit together?

In seeking acceptable evidence of an existence beyond the physical, there are glimpses that go beyond the senses. Let us picture ourselves circling with the sun around the earth. There is no more experience of night and day. Instead, our eyes make out days joining together in unbroken brightness. The closest we can come to such an experience is possibly at the poles, where for most of the year the day lasts longer and one day joins to the next with only brief periods of dusk.

Enlightenment is truly a state of self-reliance, requiring awareness of the spiritual self. It is a state of liberation from the supremacy of others over one's own thinking, but also from domination by the intellect. Spiritual indolence is often masked by excessive intellectual zeal, and with a demand for rationality in matters that go beyond competence of the intellect to examine unaided. Pure reason ought to be understood in terms of intellectual activity directed by the spirit. In this the spirit is prominent and leading, while the intellect is the vehicle for the implementation of its will.

The mind–body puzzle is as old as antiquity. The supposition that the brain is the seat of the mind dates back two and a half thousand years to the Greek physicians and philosophers. While mind and body are closely related, each reacting to stimuli and to each other, the brain may be observed and scanned, and one day it may well be possible to transplant it from one body to another. What a Pandora's box of gargantuan ethical dilemmas this would open up! Bits can even be removed and edges sutured in brain surgery. The mind, on the other hand, is free-floating and strongly bound to personality and ego. It is conscious of itself, as illustrated in such common phases as 'To my mind it is wrong', or 'The triumph of mind over matter'.

Clearly, then, our mind and our sense of self go beyond the physical. Although the 'I' appears to begin as a tiny embryo, a microscopic bunch of cells that gives rise to all the parts of the body, the development of mind cannot be traced and recorded. The self and the body are of a completely different species, and consequently our self-awareness cannot evolve from the body. And since we live with past personalities like Shakespeare and Beethoven – who are as alive today as they were during their lifetime – and may draw from them inspirations that go beyond literature or sound, film and video-recordings to fulfil some need in us (to the extent that they have indelibly changed the outlook of the living), the ego would seem to have some form of life, a continuing living essence, present both before birth and after death on earth. Indeed, all acts of will, both great and small, have their effect on the whole of existence. And since the ego continues in some form after death, then it must have existed before birth. Will or mind seems to be reincarnated, reborn in the mobility of our spirit.

Being self-aware, we know when and what we are thinking and feeling. We do more than put words in order when we are thinking. Our heart knows more than we can think. We can look inwards to monitor and question our thoughts and actions, the consequences of our desires, judgements, beliefs and intentions. How we acquire this self-awareness, or whether the trillions of cells that make up the body could come up with this critical sense of self, is a mystery. We are also not sure how reliable and stable this sense of self actually is, when at any one moment the ego might be boldly storming the heavens and the next it may be cowed down with a whimper and feeling worthless, one moment godlike, the next the quintessence of dust!

We do not know how consciousness evolved. We do not know the interplay between brain activity and self-awareness, the connections between subjective and objective truth, or between perception and reality. We know that we can act and think in ways that no other species has managed, but we do not know why this is so. All we know is that we must keep asking these questions. Will we forever be denied an answer or are we simply looking in the wrong place? Such seems to be the mixed blessing of self-awareness.

Shifting cold ashes around

Personality is at the centre of awareness, of our ability to think, to make decisions, to remember and to act, but it is inaccessible to analysis. It remains a mystery, an enigma, in spite of the attempts by modern psychology to unravel it.

While we know a good deal about the formation and activity of the brain, the nature of our mind, consciousness and personality is a puzzle, as is our being an experiencing ego, capable of finding beauty and wonder in ourselves and in the world around us. Visual experience, for instance, is not a simple replica of the visible world, not just a sensory perception caused by light striking the retina and then being processed by the brain, not a mere photographic image, but something more. No two people see or experience in exactly the same way, not even identical twins.

At birth the vacant brain is a brand new tool that will be supplied with information and continually conditioned as life unfolds. The information is received through the senses and then stored in memory, from where the intellect sets to work on the information through conscious and unconscious association. This information will necessarily be different between one individual and another by virtue of education and upbringing, nationality, culture, profession, friends and other associations, and the dispositions favoured.

Brain activity underscores cognitive performance as well as emotional experience. There is a chemistry in love and hate. Neuroscience demonstrates that consciousness correlates with neural activities in the brain. Brain-imaging technology can trace mechanisms of awareness to different neural parts, but correlation does not imply causation. The brain's network of hundreds of billions of neurons receives and gives out signals in the course of our interactions with the world. Millions of nerve strands sprout from the brain, in groups adapted to survey diverse energetic states of the world about us, relating to heat, light, force and chemical compositions. Our sensations encompass touch, temperature, pain, smell, sound, taste, light, the activity of our muscles, tendons and joints, as well as pressure, gravity and acceleration, and other fluctuations in the external and internal environments. Impulses stream to the brain where, together with signals added from our reservoir of experiences, they form the basis for our judgements and actions as well as for our appreciation of the world in and around us.

According to the experts, the brain has zones of responsibility. Modules of neurons interact as centres for memory and stored experience, learned associations, ideas, vision, hearing, language, and so on. The brain processes incoming information into neural maps that are continually updated. But even though, at the level of sensation, each human brain receives similar impressions of the world around, it will store and combine them in unique ways.

Brain-scanning devices have identified some forty or more zones in the cerebral cortex: there are zones for vision, for hearing, for movement of the limbs, for speech, for reading, for writing and so on, and the zones are connected by various pathways. Destruction of any of these zones results in the loss of the associated capacity. For example, after a stroke the speech zone may be destroyed. Destroying some parts of the brain will impair memory, rational thought, sight or physical coordination. Destroying the frontal lobe disconnects the capacity to feel emotion. The two million or so neuronal modules (distinct groupings of neurons in the cortex), each with up to ten thousand neurons buzzing with electrochemical activity, build into a sprawling network of layered interactions across the cerebral mass. Brain injuries and operations can alter the personality severely enough to suggest that the conscious sense of self arises in the interaction of any number of neuronal modules. These are switched on and off to mirror our conscious perceptions and mental states, to enact and regulate thought and emotion, and to elaborate action and behaviour. It may seem as if our brain is our mind, and our feelings simply the incessant chattering of neurons. Do the neuronal modules control our behaviour? And do they act before 'we' react with a conscious decision?

Our experience of the world, or of the meaning we make of it, is strongly influenced by what we see. The brain receives a lot of its information through the eyes. In visual perception the brain identifies patterns such as lines, edges, depth, angles, squares, stars, form, movement and other simple geometric orientations and colours. Neurons in parts of the brain associated with visual mechanisms exchange information pulses with other brain parts. Visual experience, however, is more than the illumination on the retina and the information registered by the brain. What the brain does amid the cues that come incessantly at us is to sift through the reservoir of experiences, memory, learned associations, introspection and feelings on the way to conscious perception, thus becoming part of our knowledge to inform and guide our actions and behaviour. What we 'see' is not a schematic representation of the world etched onto the retina and reflected in the mind, but part of personal experience. We see what we are *prepared* to see.

The brain may enable us to see; however, it is the *will* that makes us take notice. Thus we have the subjective nature of vision, for example. We see through the prism of what we know and believe. We are not passive recorders, respondents to stimuli, for we have the capacity to love, hate, fear or envy. We also are able to show forgiveness and exhibit compassion, we perceive humility and can express gratitude. These are not qualities we feel in our head but the special peculiarities of who we are, as are free will and the capacity to perceive intuitively. The internal state of an individual's brain tells us nothing of the personality of the incumbent. No zones of the brain can uniquely house such intimately human feelings as love, hate, joy, sorrow, wonder or disgust.

Disrupting a zone or damage to associated neural network may impair the pertinent function, but many questions arise in this respect. Nature in its unfathomed way seems at times to overrule the purely material and physical, as if, when the chips are down mind really does control matter. Destruction of the cerebral speech centre, for example, should result in speech loss. Other loss or trauma, as the result of accident or stroke, infection or an inherited disorder, can cause cognitive and motor disability and visual and hearing impairment. Degraded performance by the cerebellum causes disturbances in coordination, especially fine movements, and behaviour and development disorders. Yet it happens that some people with loss of speech due to a damaged speech centre, may suddenly speak again. Uttering a few words may be attributed to shock or sudden strong emotion. How is such a thing possible if the knocked-out brain centre is the cause of the speech disability? Clearly, something else is able to control brain activity and behaviour. The personality and other changes caused by brain damage merely disrupt the mental portal between the soul and the brain and interdict the spirit's ability to communicate with the world.

The brain reorganizes itself after loss or injury. This is known as neural plasticity and it is more pronounced in children, whereby activity is transferred across the cerebral hemispheres. Damaged brain areas in children are more easily reorganized compared to those of adults and new brain centres may take over functions from those that have been damaged. Even in adults, lost speech may be regained or paralysis freed by the repair or replacement of a brain centre. And with someone who regains his mother tongue and, in addition, acquires a foreign accent or even a second language for no known reason within a few months, it is more than memory being reconstructed; will is being imposed on brain activity.

The child's capacity to, as it were, remodel the brain and repair damage might seem inconsistent, because young children are considered to have no self-awareness and cannot therefore by self-will alone bring about greater neural plasticity. Children do not come to the world mature, yet they are not born with an empty slate for their mind either; but they are less self-conscious than adults. They need at first support and protection, and a guiding hand. They learn at a faster pace and they learn by imitation. A lack of self-awareness does not confine a child to a passive response to experience or to the equivalent of a mimetic animal. Childhood psychological foundations take in the cognitive feats of millions of years of our development from our hominid ancestors. They reiterate the stages all the way to the upright creature eager for self-development.

Each child illustrates, as in time-lapse photography, the shaping, strengthening and culmination of the self, embracing millions of years of development. The early stages of evolution covering various species are recapitulated in a compressed trajectory in the developmental stages of the human embryo, where features like gills, tail-folds, webbed digits and skin fur can be seen. The newborn encounters the world about him through sound and touch. He soon discovers his hands are for grasping and starts exploring by putting things in his mouth. Within months he has acquired simple speech and an upright gait. His initial language is without self-recognition, but in something more than two years his speech is transformed into awareness of his separateness from the surrounding world, and finally self-awareness, the uniquely human step that took many thousands of years of development to attain.

The first years of life are vital for the development of the personality and the knowledge of self. This early period is decisive for acquiring social competence and the ability to form and maintain adult relationships. For example, emotional connection and security is shaped by experience from earliest childhood, and ideally depends on a steady parental or adult connection in the first few years. After birth, sensory mechanisms linked to the brain process experience throughout life and are essential for the development of self-awareness. The brain changes throughout life in relation to experience. Children need a free rein to discover, try out and get to know the world themselves as an initial step toward the free development of a stable sense of self and independent volition.

Before adolescence the child will naturally be asking the same questions that humans have always asked and, with expanding consciousness, must continue to ask: 'Why am I here and how did I come to be here?', 'What is the purpose of life and, if there is any purpose, how can it be known?', and above all 'How do I fit into the scheme of things in this world of wonder?'.

Transparent mystery

How consciousness arises in a physical world is a giant piece of nature's puzzle. One may well wonder how the exchanges of fragile nerve cells in electrochemical muttering can make us who we are. The brain is continually moulded, reconfigured for learning throughout life, but what does the brain know of love, beauty and truth? Not to mention their opposites? The brain is a beautiful and complex learning machine, but it does not know itself to be that, because the brain is in essence neutral; it can respond to the subtlest stimuli, but not really 'know', at least not in the sense of understanding, sympathizing or having values.

We retain experience in memory. Sometimes we remember things without thinking, such as when a truth enters our consciousness unbidden, while other memories are tapped by thought processes. Long-term memory sets in for the child at the age of five to six years. It takes this long for the child's brain to develop emotional memory. As adults we consciously recall little or nothing of our first few years. Memory storage entails structural changes at the neuronal connections; day-to-day memory is stabilized in strengthened signals in the synapses – connections between neurons – of critical neural circuits. The mind is continually buzzing with events throughout life, and new synapses are built, some to replace others. In other words, our views change, and the cards are, so to speak, constantly being re-shuffled.

The question here is whether neuronal centres initiate action and control behaviour before we respond with a conscious decision? Is my finger going up or am I raising my finger? Is the brain responsible or is it the will behind the brain that must bear the ultimate responsibility? Should we credit the car or the driver? The body needs the soul just as the car needs the driver. A car without the driver is as good as a pile of scrap. The driver, however, is independent of the car, can change to another car in the case of a breakdown, or use other vehicles of mobility. However, misuse by the driver can destroy both the car and the driver.

We are generally rational with decisions that matter deeply to us. Yet we may make many of life's big decisions based on our intuitions, thus without rational analysis. For example, we may let our intuition guide our thoughts in deciding to marry a certain person or choosing a certain career path. We react to certain perceptions automatically, one might almost say mechanically, seemingly without conscious choice. We act without thinking, and react spontaneously to a threatening situation before we have had time to think about it. We often have no idea why we do many of the things we do, and are often unaware of why we feel as we do. We do not remember many things we say and do. Or if we do, the memories are vague or distorted. What we remember or wish to remember leaves gaps open. Memory does not function like a video-tape replay or remain like fixed deposits of experiences in a memory bank; it is reconstitutive, and rather like wandering in uncharted territory and back, ambling along an imprecise route. Things look different in hindsight. Memory is imprecise and impressionistic. Loosen or tighten a thread here or there in the perceptual web and a different coherence emerges. No two observers retain the same perception, and each will give his own impression. The brain is not a facility of enduring memory.

While thoughts are generated in the brain, the cerebral sprawl contains a limited reservoir of experience. The intellect is liable to fragment and distort memory from being an actual recall, and condenses or downgrades higher emotions like joy to pleasure and love to sensual feeling. It stokes the fires of passions to great effect.

The 'small' back brain, the cerebellum, offers the portal to our great reservoir of influential experiences, most of which do not surface to the light of day-consciousness. The unconscious mind is for the most part inaccessible or accessed mainly in dreams or states of sharpened awareness. Unconscious psychic processes, powerful auxiliaries of ego-consciousness, play a more or less unknown role in the waking state. The unconscious is studied in depth by psychologists, often in the treatment of the mentally disturbed, occasionally revealing those unsuspected chasms in our psyche that produce 'abnormal' or psychotic behaviour. We are at a loss to know how much of our behaviour is conscious and how much is unconscious. The subliminal unconscious has been exploited to good commercial effect by Gothic writers, surrealist artists, film directors and marketing gurus. Be that as it may, it attests to that undeveloped intuitive self which resides in the soul.

The soul has, in an age of scepticism, suffered by neglect. It is our enveloped spirit core, acting through matter yet essentially alien to it. The spirit is implanted or incarnated in matter for a time, but it remains fundamentally different from it.

Intelligence is a much-prized quality, and rightly so. It is a manifest indication of the ability to perceive, understand and reason, that is, to draw logical conclusions from the evidence at hand. A functionally normal brain will underscore intelligence; but this may be undermined by indulgence in alcohol and drug abuse and by disorders of the brain that cloud self-recognition. On the other hand, intelligence is not always the key to wisdom or to the positive development of our species. If used for destructive or selfish purposes, intelligence will set humanity at war with itself and hinder development for perhaps a century or more. It can wipe out entire species, including our own.

It is more the inner qualities, the intuitive knowledge of our spirituality that spurs us to humility and inspires us to love and understand others. Spirituality does not inspire one to hate and despise; these are the artefacts of narrow-mindedness, the result of the intellectual rather than the spirit's domination of the mental landscape of the mind.

Physical disability has often proved no bar to miraculous intuitive feats. No-one can afford to look down on another; no-one is so intelligent that he cannot learn even from anyone who is physically or mentally inferior to him, in as much that such unfortunate individuals may possess spiritual qualities, including spiritual intelligence, that far outweigh their physical deficiencies. All human beings have individual lives that have meaning, unless they themselves make it meaningless, and each one must struggle in his quest for inner spiritual development.

While it is true that genes affect mental ability, intelligence and cognitive skills are no more governed by genes than self-awareness is under the control of the various regions or neural networks of the brain. Genes do influence intelligence, however, playing a passive, implementing role in behaviour, but the spirit of a person overrides the interactions of the neuronal circuits. We can use our spiritual resources to train our intellect into subordination to intuitive recognition.

Intuition is fundamentally of more value than rational intelligence. Not to understand time is a disadvantage, but not to know when to act could be disastrous. Intuition, and not the recording brain, will tell us the best time to act and to deal more effectively with happenings. For the brain is but the tool of the spirit. The spirit's activities do not coalesce with our perceptions and memories within any particular structure or part of the brain that engenders the sense of self and thereby establishes a physical basis of personality. The mind is not 'sitting' in our brain, with our unconscious mind transcending it. Our mental life is anchored in the invisible core of spirit, which animates the body that is to be the earthly implement of its expressive power and it merely makes use of the brain.

The self that endures

Our large brain, with our intelligence and complex behaviour attributed to it, is regarded as the means by which we achieved domination over other species despite an insignificant beginning to our evolutionary story and our relatively brief history. Millions of years ago, our ancestors were small tree-dwelling primates. Our species followed upon the zenith of ape evolution, which had achieved the heights of the development of the species. In the movement of evolution, most species become extinct. But the adaptation to different environments can open the way to new species, for instance, in the case of the dinosaur, from reptile to bird.

It took huge tracts of time for the planet earth to condense out of cosmic nebulae some four and a half billion years ago, and in the earliest stages it was without life forms. Hydrogen compounds provided the conditions for the appearance about three and a half billion years ago of simple bacteria-like forms. Plant life gained the first foothold with the simplest, least demanding lichens, mosses and algae, which needed but a modicum of light and water. Then followed ferns and flowering plants, to the point where conditions were ripe for feeding by land-colonizing animals. The most complex life forms have emerged essentially in the last six hundred million years. From the primeval sea there emerged the simplest organisms – which in truth were by no means simple – and these were transformed through fish and amphibian stages to land-dwelling species. The sequence lay in readiness in the all-embracing power in which the cosmos swings, a pattern emerging by degrees, with each step in the life process finely graduated.

Maturity of a species opens the possibility of a 'next step', a gigantic leap that may take millions of years. The reptile began transformation towards a bird, and at the zenith of animal development the ape body opened the way for spirit incarnation. This opened the way for admittance of our species of spiritual being into the highest developed animal body in the form of the hominid ancestor. Not that our spiritual species evolved from the ape-like ancestor; the spirit is not only a fundamentally different species but it has nothing in common with the core of the animal and could therefore not evolve from it. Humans took on further development of the animal body to greater ennoblement by virtue of their spirit core.

Animal souls have their origin in the animistic planes, which lie lower than the spiritual planes. The spirit germ of humankind already contains, in nascent form, everything he needs to develop self-consciousness. With the animal, in contrast, the soul must first form. Animistic accumulations incarnate in splits in animal bodies, akin to entanglement of quantum entities. The incarnation into a matching body is marked by an impulse for formation of the soul. The animal usually loses its form after the detachment from the physical body; it is reabsorbed by the homogeneous band, the so-called 'group soul'. The earthly development brings increased warmth to the homogeneous whole. We humans play a role in as much as the animal can, through a bond of love with a human, be highly energized, receiving a supply of power that binds the animal soul more firmly, such that it retains its form beyond the earth.

Our appearance on earth marked the culmination along the milestones of evolution, especially the changes in primate development. Our conscious mind suggests that there is no continuity between body and spiritual attributes. Genealogy apart, we are obviously much more than an animal species striving for survival, competing for scarce resources, or a tool in the whip-hand of blind nature.

The biblical story of creation would appear to be incongruous with science with regard to the evolutionary trajectory to humans. The phrase 'God created man in His image' implies that humans are imbued with something other that distinguishes them from beast and plant. Whatever our feelings about the naivety of the Genesis story, the lesson it teaches contains an enduring truth about human nature. It is surely no mere hubris to say that the truly living part of a person is not his body, not even his highly developed and much-prized intellect, but his spirit.

Some three million years ago, almost yesterday in geological terms, climatic transformation of fertile African forest into scrubland and savannah introduced a new life form, the hominid, the peak of animal development. The material vessel was in place for a totally new and fundamentally different species, a creature that could attain unparalleled cognitive feats. The species understood things other creatures could not, asked questions, and acquired language and advanced technical skills.

The first humans lived in small cooperative bands, grew alienated by their nature from even the highly developed animals, sheltered in caves and by rocky surroundings, with herd instinct and the cave providing warmth and protection from predators. Using their upright stance to free the upper limbs to hunt and gather they grew adept in the use of tools.

The earliest humans are acknowledged as barely superior to apes: similar but different; close but in so many ways vast distances apart. Thus the upright gait and skull shape, as well as the descent of the larynx that enabled the development of speech are manifestly human characteristics not shared by other species. Above all, we have free will, a far-reaching perception beyond material concepts of time and space, and can conceive of an ideal or perfect form. And that is no small wonder. The human spirit has unfolded from an unconscious germ toward a self-aware personality. Spirit is different from the animistic soul of animals. The notion that we are little more than highly complex thinking apes, perched at the top of the tree of life, is like saying that the computer is a descendant of the abacus. Unfortunately this view is shared by many people today and, is in part, derived from a narrow interpretation of Darwinism as well as from a total ignorance of the existence of the spirit as the seat of human consciousness.

A mind that watches and receives

The computer is a marvellous invention, one that has changed our lives and the way we view ourselves. But, while a vast improvement on the abacus, it has its limitations, as we all know. The computer, with its integrated circuits, can execute an astronomical load of logical and mathematical instructions in a fraction of a second. It far surpasses the human brain in analytical power and speed. Neural impulses travel at a snail's pace by comparison with the computer's internal messages. While a computer can perform narrow functions with lightning speed, the microchip processor has little flexibility. The human brain, in contrast, is impressively flexible, with parallel routes of vast neural networks enabling it to integrate the myriad impressions feeding the mind with instant recognition.

In the developing brain of the human embryo a mass of neurons are generated, perhaps twice as many as necessary, to cater for its exploration of the world and to nourish its perceptual abilities, which go on developing during the first months of life. In our early years of rapid cognitive development, the brain processes a deluge of experiences; neurons proliferate and certain interconnections between groups of neurons are strengthened at the expense of others that are little utilized. It is true that from a certain standpoint the brain functions in a similar way to the machine code of a computer. Both are instruments, although of different ranking; the brain's job is to serve the spirit, and behind the computer is the brain, whose invention it is. The computer is an electronic brain of sorts, but it can do nothing without a supply of data. It needs programming. It can then carry out many brain functions, as we have seen, far more efficiently than a brain – a case of the pupil outstripping the master, which is the case with all tools. We would not need them if we could do better with our bare hands. But a computer cannot produce anything new, can initiate nothing; it can only combine known factors.

Similarly for our brain, information reaches its 'databank' from experience and tutoring, followed by evaluation of the data. The brain, much less efficiently than the computer, sorts, classifies, stores and retrieves information. But here the parallel ends. The brain, except when its activity ceases at death, is never turned off in its power to think and feel. It is capable of forgetting, misapplying, distorting and misremembering. It seems to be rather an inefficient machine, perhaps like an overgrown forest. Laden with emotion it can wrongly sort, misapply, get fatigued and simply give up. Unlike the computer, it has 'a will of its own'.

The brain works 'hand in glove', as it were, with the emotions and the imagination. A computer may do impressive work, but it cannot love its work, cannot support an ideal, and has no sense of responsibility. It can never bootstrap its activity to a sense of beauty or justice, and never know good from bad or right from wrong. Just as it cannot let itself be inspired or have sudden intuitions about the way a problem may be solved. Of course, on the positive side, it cannot get impatient, or become annoyed about its workload; neither can it refuse to do what it is asked. The computer is a brain without a mind. The self-conscious mind, on the other hand, is the interface between the brain and the spirit. We have an intuitive mind and a rational mind. Mind and body form a certain union, when we feel pain, for instance. Every mental state is not necessarily a brain state primarily. Mind expresses the power of the spirit's volition in thought, language, choice and action. It makes possible our introspection, yearning, self-identification, creativity and sense of beauty and justice, as well as alertness to the converse, mindless boredom, ugliness and injustice. It is mind rather than brain that is at work when we show courage, endurance and perseverance. The brain is the highly flexible tool at the mind's service, but the brain does not initiate anything of enduring worth. In this respect it is not unlike the computer.

Physical sensation, intellectual process and imagination are brain states. Intuition stirs from our innermost core of spirit. Its emotionally charged animation crosses through the bridge of the solar plexus, situated in close proximity to the pit of the stomach and sends dense neural projections to connect with all parts of the body. Sensitive people feel the behaviour patterns of the spirit in the region of the heart or stomach. The solar plexus is a network where nerves that usually work quite separately come together and communicate in a small central exchange. It forms part of the autonomic nervous system, an auxiliary system of involuntary regulation.

This autonomic nervous system belongs to an ancient autonomous pathway, one that was predominant before the evolution of the neocortex. It is dominated by the brain's primitive structures like the brainstem and the hypothalamus, prime centres of the hard-wired alarm system in lower vertebrates. Its neural pathway, for example, mobilizes our reflex behaviour patterns in tense situations. Its counterpart, the *central* nervous system – with its centres of thought, imagination and planning – is dominated by the higher centres such as the cerebral cortex, thalamus and the cerebellum. These are evolutionary newer areas grafted upon the alarm-and-alert mode of the autonomic system and they provide the cognitive awareness and complex decision-making of the higher animals, being most developed in human beings.

The solar plexus is exceptionally configured to receive and transmit sensory information rather than to process it. But it does not receive impulses from the viscera of the intestines. Sensation in the hollow gut runs along nerve pathways to the brain, bypassing the solar plexus. Ninety percent of nerve connections in the stomach run straight to the brain. Most neural activity in the abdomen is the result of local autonomic reflexes and is not often felt. Despite this, the solar plexus is stirred by impulses not from the viscera but, as it were, from beyond the body, from some unfathomable depth 'inside'. Thus, on a physical level, people who have breathing problems may find relief through massage of the region, even though the area plays no part in breathing. On a psychological level, mental blocks or fits of 'nerves' may be eased by the application of a compress to the solar plexus. It is the area where we often find the capacity to relax, to let go of all those seemingly vital concerns that the brain hugs so tightly to itself. Deep breathing is not merely a fashionable cult, but a tapping into the spirit world.

According to *In the Light of Truth: The Grail Message* by Abd-ru-shin, the solar plexus is designed to receive information from our immaterial core of spirit. It is supplied by trunks of nerve fibres projecting from the brainstem, the stalk on which the cerebrum and the cerebellum are perched. Direct communication is thus enabled between the solar plexus and the brain. Studies can identify the neural networks that process our emotional responses – love, hate, fear, joy – and their links with our thoughts, moods, memory and behaviour. For example, a situation of fear or excitement evokes a rapid heartbeat and sweating that mobilizes the body to respond. Yet we experience joy, love, fear and anxiety, not with our head but in the pit of our stomach. Intuition, in reality the manifest voice of the spirit, is called 'gut instinct', and at anxious moments the stomach gets 'tied in knots'. People who lack courage 'have no guts' or 'no stomach for the fight' and revulsion may cause people to 'throw up'. The emotion of pure love is linked with the heart. A person devoid of emotion is said to have a 'heart of stone', and one who suffers has a 'heavy heart'. Thus affairs of the heart may better be called affairs of the stomach.

At the solar plexus, the silver cord of fine material substance that is prominent in descriptions by clairvoyants connects body and soul. The cord, part of the astral body, enables the soul to animate the physical body and manifest consciousness through brain activity. The cord dissolves at death, and it loosens during sleep. Sensitive persons woken from sleep have even spoken of feeling their 'selves' entering into their body via this region.

Not every mental state is a brain state or even a physical state. The custom today may be to shrug off these finer experiences as 'pure imagination', as 'auto-suggestion', or as some kind of 'mumbo-jumbo'. Whether we wish to acknowledge them or not, we are constantly receiving ethereal impressions that affect our perception, thoughts, consciousness, moods, motivations, memory and behaviour through accessible and impenetrable brain processes.

In and out of the mist

An over-cultivated intellect distorts the mental state. Our ability to perceive, what we make of an experience or the way we understand the world, becomes strangled and narrowed. While the brain receives and processes the ceaseless sensory flux, the spirit moves by intuitive perception, that is, by pictures, unmediated by words or rational analysis; thus hunches precede rational understanding. The brain will interpret practical everyday matters, such as where to find food, shelter and a mate, or how to survive in a fierce or at least in a fiercely competitive world. The part of us that feels deeply, that yearns to be useful and is impelled to ask unanswerable questions – at least by the intellect – that conjures up new worlds out of air is, in other words, the part that is the soul, or more precisely spirit, and it is subject to different vibrations. It marches to a different tune to that of the intellect. There is the bound volition of the intellect, bound as it is to the earthly, above which, despite its tremendous achievements, it can never rise, and there is the free will of the human spirit with its ability to soar above the earthly concepts of space and time and thereby to fathom the 'hidden connections' that determine our earth-life.

To begin to understand the puzzle of human conscious awareness we must first investigate the activity of the uncelebrated cerebellum. It is a cauliflower-shaped structure at the back of the brainstem; it is known to function in the control of movements, for instance fine coordination in walking or driving. While the cerebral cortex is the domain of intellect and imagination, the cerebellum belongs to intuition. Only the cerebellum can absorb spiritual impressions. Intuition, as a wave of power from the spirit, stirs through the solar plexus to impress a certain volition upon the cerebellum; the impression is transformed and is immediately transmitted to the frontal brain, from where the volition may issue in thought, language and action, thus the effortless intuitive perceptions or gut reactions. Intellectual activity eases our life on earth; the cerebellum carries intuitive perceptions that are timeless and spiritual, messages, as some might say, from the Creator.

When we need to remember something we say we will keep it at the 'back of our mind', which is where the cerebellum is. In reality, how deeply we perceive something depends how strongly an impression is made on the spirit. The implicit memory or ingrained experience can be recalled across time and space. The same applies to the expression 'learning by heart'. Incoming sensory information processed through the frontal brain, the intellectual instrument, is passed on to the cerebellum, the intuitional brain, and via the solar plexus it is absorbed by the spirit. This is the reverse of the pathway followed in the expression of the spirit's volition, which has the quality of immediate certainty of intuitive perception. Often we feel intuitively urged to make a certain choice or take a certain action. The exchange between sensory impressions and spiritual promptings is central to our distinct capability of self-awareness. Beside the stream of sensory information, the brain also receives impressions from the spirit.

Our consciousness is an expressive attribute of the spirit. The spirit is nowadays identified with the body and synonymous with brain functions, but how may subjective experience, which characterizes our self-awareness and is intrinsic to our personality, be correlated with objectively measurable brain events? As already outlined, the spirit needs the brain but is itself not 'of the body'; it is of a different consistency. It uses the body as an implement and the brain as a tool to express its volition during its sojourn on earth. Sensory impressions are evaluated and indexed by the brain – that is, they are sorted through memory. The brain goes on to resolve what manifests explicitly in consciousness. In the process, the concept of the mind is an interface between the brain and the conscious spirit. It is a dynamic filter that is subject to the influence of both spirit and brain. For example, our spirit has not only a great influence upon the body but it can exert a decisive influence to focus our conscious attention on a particular happening or issue out of a stream of impressions.

An indolent spirit is seen in a person who submits one-sidedly to the intellect. The brain then has the greater influence on the filter of the mind, and much of the weakened information from the constrained spirit does not penetrate to consciousness or does so only in fleeting, intractable perceptions. It is as if such a person is two separate minds, the intellect seemingly with a mind of its own, and the resourceful activity of the spirit virtually hidden or shut off. The nature of the spirit is beyond comprehension by the brain. Our self-consciousness is not completely realized in the natural limitation of the intellect to earthly concepts of time and space, otherwise it becomes hobbled, in a total reality comprised of gross matter. Though the spirit is superior to the brain by nature of its higher and finer species, in the case of spiritual indolence, perception is mostly limited to the brain and intellect, and withheld from the spirit. This is the foremost conceit of the devotee of reason. This situation also has its parallels in business life. If a manager is lazy and takes his eye off the ball, so to speak, then his staff will only feed him with information that is favourable to them – or none at all. He eventually becomes irrelevant and is then replaced, unless he brings the whole organization down with him first.

The simplest movements adjust to learned patterns in the brain. Learning requires a conscious effort of will. After sufficient practice, the will is relieved of further spiritual impulse in the automated brain functions. As an example of this, a person only learns how to walk, swim or ride a bicycle once, usually as a child. Once he has mastered these skills he will never forget them and they will be at his disposal for his entire lifetime. Even if he has not used these skills for some time, say, in the case of the bicycle, they will 'come back' to him with minimal effort. In one sense, the will is freed of baser concerns and can then be exerted to set finer goals or values. Spiritual will is then able to stretch the skills of the brain to higher abilities; the superior spirit needs always to expand the brain's learning processes by exercising its free will. This can again be illustrated with language skills and artistic pursuits. Many people learn a foreign language or play a musical instrument in their youth. If they are serious about their pursuit, the spirit is also involved because the spirit can express itself both in art and in language and they become quite adept. However, in contrast to simple motor reactions and skills, these skills are greatly diminished or even totally lost in later life if they are not constantly practised, but this again is the result of a choice the spirit makes, perhaps because other matters are now deemed more important and take up too much time. Without such freedom of choice, the activity of the brain is soon largely exhausted in repetitions.

Indolence of spirit means repetitive learning dictated by the brain and it is characterized by a rigidity of perception and outlook. Without guidance by the spirit, perception is filtered by the brain to exclude all that contradicts past thinking habits and expectations. Doubts arise with the unusual and the unfamiliar, from which without spiritual impulse the brain refrains, because the effort would mean overcoming contradictions and require new efforts in learning. We become adept at interpreting the world to fit in with our existing beliefs. As spirit, we can have greater freedom of self-expression as well as greater influence on the mind, and not be subject so extensively to the habits and superficial desires of the brain. Without exertion of the spirit, any ability to recognize the intuitive deficiency is prevented by the dominance of the brain together with its intellect over the landscapes of the mind.

Experiences from a former existence on earth cannot be found in time-conscious explicit memory encoded in the brain. The brain decays after death and the new brain at birth is yet to be impressed with information from the senses. Perceptions imprinted on our spirit become part of our personality, character and behaviour, in the abilities and talents we bring along as developments from our wanderings. Memory is important in unravelling the mystery of consciousness. Indeed, without it there would be no consciousness, which we depend on for sorting, filtering, storing and evaluating impressions to guide decisions in our everyday lives. Clearly, if there is no memory, there can be no awareness of perception or thought, and no possibility of foresight and the exercise of will.

The cerebellum has far fewer neurons than the highly developed cerebral cortical mass, yet it may have an equal number of neuronal connections. It has extensive connections with all structures of the brain, but its signals are not very vibrant because of its stunted development. Although some thinkers have also come to the conclusion that there was 'something amiss' with the human brain, they could not pinpoint the problem. It is only in the *Grail Message* – and nowhere else – that we find a comprehensive explanation of what is actually *wrong* with the human brain, how it came about and why, and the implications for humankind as well as what we have to do to rectify it. The *Grail Message* gives the biblical term 'hereditary sin' (an expression nobody was able to do much with before publication of the book) a whole new meaning in fact, by linking it with the imbalance that exists between the two parts of the brain – that is, the cerebrum and the cerebellum. We caused this imbalance ourselves by wanton interference in the natural order of things, through an excessive cultivation of the intellect at the expense of greater inner mobility, and this has now become inheritable in the effect of the laws of nature that govern the mechanism of physical heredity.

Stated briefly, our large cerebral brain has developed too quickly, evolved at too fast a rate, for our own good, but as the *Grail Message* explains, this was caused by ourselves due to our voluntary decision to value intellect over spirit. This has happened at the expense of the intuitional brain, which has become stultified through our lethargy of spirit. The excessively developed intellect now smothers a weak intuitive perception, narrowing comprehension and inhibiting insight.

The intellect can only provide a peep-show stage that reduces the magnitude of reality. Untrammelled empirical logic clouds the insight that the physical reality is a coarse replica of an ideal world. The greatest thinkers, poets and philosophers have always felt the presence of the spirit, which is merely reflected in the material world. Human beings are vessels in which sparks of creative spirit dwell for a time in matter in the course of personal, self-conscious development.

Dreams of being awake

The brain operates by a code of communication modulated by changes in the strength of the contacts between neurons. Neurons communicate by electrochemical impulses. One neuron makes thousands of contacts with others, and these dense interconnections may link millions of nerve cells in changing alliances. The neuron is integrated into neural networks that are related to particular learning and memory tasks or mental states.

The frontal brain has been studied in some detail, but the cerebellum, with its deep centres, is more difficult to observe and measure. Present recording techniques and brain scans are not adequate enough to measure the much faster impulses of the finer nature of intuitive perception before the impulses penetrate the cerebral connections, thereby increasing their density. Such studies that have been made, in sleep therapy or dream interpretation, for example, will inevitably be tentative and inconclusive, for how can we know anything of the spiritual and immaterial that cannot be fathomed by rational thought alone, since the frontal brain, upon which we rely so much, is not equipped to do so, and the back brain is not up to the task due to its weakened state.

From the moment we fall asleep, neural activity begins moving from the frontal part of the brain to the back, to the cerebellum. Waking reverses this process, with electrical activity spreading from the back of the brain to the cerebral sprawl at the front. Sleeplessness and disturbed sleep result at times in nightmares and somnambulism, which are of course not especially modern phenomena. However, the reliance on drugs for a peaceful slumber has escalated in recent times. There was a time when dreams brought precious advice and warnings. One might truly have said of a particular problem to 'let's sleep on it' and have expected to find a resolution. Most of our troubles can be directly attributed to the overworking of the cerebral part of the brain.

The cerebral brain and the cerebellum should be balanced in development, perhaps be about the same size, then each part would complement the other and contribute uniquely to the harmony of the whole, which is necessary for depth of perception. But this has not happened, and an over-reliance on the frontal brain, our rational tool, devoid of intuitive input, has led to warped and damaged lives, and indeed shrivelled souls. For the intellect can plainly be 'stupid' as Swift shows us in *Gulliver's Travels* and, in our time, we have witnessed (at the cost of many millions of lives and dwindled souls) sundry futile attempts to re-model the world on theories and ideologies of various kinds. These have collapsed in chaos, leaving untold suffering in their wake. They have been replaced by new ones, which are often the old ones in a new guise. Only one example will be cited here. We have 'globalization' instead of 'imperialism', the imperial powers of yesteryear being replaced by mega-corporations who are often more powerful than the governments of the countries they operate in. They keep millions of their citizens toiling in virtual slavery and, supported by corrupt officials, ruthlessly appropriate their resources and destroy their environments so that only a select few can live a life of luxury on the back of rising share prices and outrageous salaries. What is the difference between this and the situation that existed with the rise of imperialism and colonial expansion in the eighteenth and nineteenth centuries? Inevitably, the harvest must come, and any rotten fruits are already beginning to show. It is as though the intellect is in a rut from which it cannot escape, making the same mistakes over and over again.

To right the human psyche the brain must evolve in proper balance, thereby enabling these two important complementary parts to work in harmony and so allow us to transcend our hobbled condition, so that, in time, the brain will recover fully from its abnormal state. Paying heed to intuition opens us to a fulfilled life of true happiness. Without it, thoughts and volition swirl only in recycled forms, bringing uncertainty as to what life is all about.

Intuition is an immediate recognition made from our spirit core, which has transcendent survey in space and time, and is not a perception exclusively informed by reason, explicit memories and learned associations. Some snap judgements and courses of action impress us inexplicably, sometimes after we have pondered long on a matter. For example, we form instant, unreasoned positive or negative impressions of people and certain circumstances. We often experience an inexplicable fondness for one person and an irrational dislike of another. These perceptions are not without foundation. In pure intuition, the spirit sets a decision. It is communicated by a wave of power via the solar plexus to the cerebellum, then it is adapted for the cerebral mechanism, where the rational perspective of the intellectual process finalizes an appropriate expression, which might be a thought, a word, an action or an impression for future behaviour. Our unconscious intuitive and conscious rational minds unite to make possible the greatest of cognitive feats that must astonish our 'thinking brain'.

The intuitive mind is nurtured by silence, contemplation and committed volition, and striving for the good, the good not just of ourselves or some abstraction called 'humanity' or 'the environment', but what we know to be good in *itself* because the spirit intuitively tells us it is so. The spirit knows what is 'good' because it originates in a part of Creation where that which we call 'the good' is embedded in life itself. Left to itself, the spirit will always choose the way upwards, towards its origin. It is only the intellect, if it is allowed to dominate, that holds it down. Thus we should – as the *Grail Message* exhorts us – let the spirit within us 'have its way'.

Lifeblood of flesh

Humankind is enlivened by the spirit; we are as alive as the degree of our inner exertion. An alert spirit, keenly engaged in life, radiates totally differently to someone with a spirit shut within, without a high goal in life, who only draws upon the basal energy of his slumbering spirit to exist, no matter how great his or her intellectual stimulation or acumen, or how high the goal of the intellect itself. This is exemplified by various infamous dictators and despots throughout history, all of whom had 'a high goal in life'. The bare intellect is a most inarticulate and limiting tool, which, at its worst, engenders folly and catastrophe.

All our mental and physical states operate on the power of the spirit. Its energy powers the ineffable and largely unconscious perceptions that guide most of our everyday lives. An enthusiastic person acting joyfully not only feels light, but also brims with energy and vigour. It is uplifting to be around happy people. A frustrated and miserable person acts without joy and feels depressed and heavy; he is lacking in mental energy and dynamism, and senses that he must expend a lot more effort than usual to overcome his inertia in thinking. It can be depressing to be around depressed people. The same activity can thus stir a lot of energy and vigour, or it can burn us out, depending on whether or not we carry it out with verve – that is, with the spirit. We draw on two types of energy in the course of our activities: energy provided by our body (for example, obtained by combustion of glucose in the muscles) and energy from our spirit core.

The spirit's energy holds body and soul together, motivates our finer aspirations, rejuvenates our bodies by warding off fatigue, too-rapid wear and tear, and unhealthy changes. Spiritual inexertion diminishes the body's strength. Consequently it tires and wears itself out quickly. Tissue regeneration is slackened and ageing sets in faster. A weak, lethargic or inactive spirit exposes the individual to physical and mental ill-health.

Blood streams throughout the body, supplying oxygen and nutrients. It removes waste matter, transports hormone messages, maintains electrolyte and fluid balance, and plays a role in the body's defence mechanism. Without blood the body is inert, for blood is the bridge through which spirit reveals itself. The body is not alive because of the blood, but because of the *spirit*. Blood actually has a radiation and this radiation is suffused with spirit, but of course it is not spirit in itself. Lifeblood can be lost (or spilt, as we say) but spirit cannot. Blood is material, spirit is immaterial. Blood and spirit form a certain union by radiation. The radiation of the blood is more rarefied than radiation from other materials. The quality of any radiation depends on the nature of the emitter, and so it is with blood. Each person's blood is different, with more than thirty different biological markers found in countless combinations. An individual's blood radiation provides an appropriate bridge for the spirit, but this blood–spirit bridge can weaken, however, through illness or other causes such as an unhealthy diet, taking drugs and alcohol abuse. If the blood radiation is somehow altered, but not to the extent that the spirit has to leave the body, then another (disembodied) spirit can 'take over' the body periodically and force it to do its bidding. Multiple personality disorders are often just that. The key to a cure is to change the blood radiation so that it only corresponds to the 'true owner'.

Pulsing in spirit and blood

Since the blood is the bridge for the activity of the spirit, it follows that any change in the quality of blood will affect the union of body and soul. The average volume of human blood is approximately six litres, but apart from exceptional situations of severe bleeding, changes in blood volume are infrequent and minimal. Low blood volume causes weakness and loss of consciousness. The lower the blood volume, the more 'distant' an individual will become. With heavy blood loss, the bond between body and spirit is loosened and the spirit exits the body. A rapid blood transfusion can restore normal blood radiation and prevent death by restoring the blood–spirit bridge. Sometimes the transfusion will consist purely of the fluid plasma, that is blood without any red or white blood cells.

Blood composition can be changed by diet, ageing or disease. Resulting changes in blood radiation in turn affect the integration between body and spirit. Advancing age is attended by weakening blood radiation until the body finally gives up the soul, that is, the radiation becomes so weak that it can no longer 'hold' the soul.

Blood composition plays a major role in the development of temperament, too. Different temperaments correspond with different age periods, like the changing seasons of life. Temperament is an attribute of the animated body, and is in part hereditary, prompting vernacular references to light, hot, heavy, or easily excitable blood. A certain radiation of a particular kind of blood composition is related to a corresponding release, primarily in the brain, and manifests a generalized feeling. In infancy, for example, the generalized feeling usually appears as a sanguine temperament – the carefree life of the moment. A young adult is absorbed by longing full of dreaminess, and has a melancholy temperament. Adulthood is choleric and is followed, in old age, by a phlegmatic long-life attrition of calm reflection.

It is worth noting here that the years of childhood, in which a sanguine temperament prevails, as well as the years of adolescence, which are characterized by a melancholy temperament, have been significantly curtailed by an education system that one-sidedly trains the intellect from a very early age at the expense of the finer stirrings of the soul, and by an unrelenting stimulation of the sexual instinct in young people, especially from the media and the entertainment industry. They have significant influence on our sensitivities, perceptions of reality and behaviours, and all in the service of commercial interests. It may well be that promiscuity among our youth is partly because their melancholy temperament is not fully awakened by the time they have to deal with these matters, which now happens at a much earlier age as a consequence of this artificial stimulation. If in full vigour, the temperament itself would act as a barrier to such behaviour, by providing a strong support for the awakening 'cautioning' sense of physical shame, which should occur at the same time. In fact, such behaviour is more in tune with the carefree sanguine temperament found at a younger age. On a more positive note, because we are living longer (in western countries at least) many of us are likely to experience the phlegmatic temperament than in the past. If engaged correctly, this temperament helps us come to terms with our eventual passing, which is but a transition.

The prevailing temperament among a people or nation, also an expression of the general homogeneity, is influenced by their geographical zone, the radiations of the stars, the kinds of foods available and other regional characteristics that affect their blood composition. Today, in an era of mass migration across the globe, both forced and voluntary, many people find themselves living in the 'wrong' geographic zone. The longing for one's homeland may also spring in part from a physical need. If one has the means and possibility to do so, one should certainly return from time to time to their homeland. The body will be invigorated thereby and gain new strength.

Sugar is the energy fuel of the body. The normal blood glucose concentration is between eighty and a hundred and twenty milligrams for every hundred millilitres of blood. An abnormally high level is called hyperglycaemia, and an abnormally low level is referred to as hypoglycaemia. Depleted glucose levels cause physical and mental impairment, and hunger is the normal symptom of a low sugar level. An abnormally low blood sugar level brings on a mania for sugary foods. As sugar levels drop, tiredness increases, thinking becomes difficult and confusion sets in. As hypoglycaemia increases, so do the symptoms and signs of faintness, headache, nervousness and extreme weakness, 'rubber' legs and an inability to think. Further deprivation may well result in personality changes, loss of consciousness, convulsions, coma and death. The blood composition is changed by an abnormally high or low sugar levels, and because the brain relies exclusively on glucose for its metabolism, consciousness is lost if the low blood sugar level impairs the radiation bridge between the body with the spirit.

A temporary high sugar level is normal after eating a meal rich in glucose. After a heavy meal we are less active and feel sleepy as blood surges to the gut to absorb food nutrients. The body lowers the sugar level to normal by quickly secreting the hormone insulin from the pancreas. In certain conditions, such as diabetes, impaired insulin secretion or insulin sensitivity causes glucose intolerance. This may mean that blood glucose levels remain high and may climb to dangerous levels after a heavy consumption of sugar, causing a hyperglycaemic crisis. If unattended this will, as with hypoglycaemia, lead to coma and death. When the union between body and spirit breaks, the radiant bond is so weakened that the bridge collapses, the body returns to matter and the soul exits, returning to the otherworldly planes in accordance with its density (in essence, the density of the cloaks surrounding the spirit). This relates to the Law of *Spiritual* Gravity.

Deficient insulin activity causes high levels of blood glucose and cells that are undernourished. The body then resorts to breaking down protein and muscle tissue or stored fat to compensate for the malnourishment. The brain is especially affected by this process. Quite apart from emaciation caused by this consumption of muscle tissue, which would be lethal in itself if left untreated, excess fatty acids in the blood due to breakdown of fats change the blood radiation drastically. If tissues cannot function properly as a consequence of the deficient nutrition, increasingly acidified blood leads to a comatose state. The changed radiation of the blood loosens the union of body and spirit, and death will eventually follow if the union is completely broken.

The body–spirit bond is also loosened if a diet is deficient in certain essential nutrients. For example, humans might have originally been adapted for a vegetarian diet, starting out from the fruit diet of apes, but over many thousands of years we have adjusted to eating meat. A strict vegetarian diet adopted temporarily can indeed rejuvenate us, but a permanent vegetarian diet would deprive the body of various essential substances. This is an issue with the vegetarian who eats no meat or fish, and the vegan who excludes all animal products from the diet. The human body today cannot suddenly adapt to a vegetarian diet, but requires transitional steps, otherwise the body and spirit are bonded by only weak radiation. Those who maintain such one-sidedness in their diet on a permanent basis often give the impression of not being fully grounded in the world. Some may, in addition, welcome their aversion to material goods as an indication of spiritual advancement and enlightened liberation from coarse instincts, but in reality they have not distanced themselves from desire – just from the *object* of desire. However, as with everything, quality counts more than quantity and moderation is the key. Thus the consumption of free-range poultry or beef or fish from unpolluted waters perhaps two or three times a week will do more for the general health than more frequent consumption of meat derived from livestock produced by intensive farming methods, where the animals are kept in cramped conditions, unable to move, and force-fed with high protein foodstuffs treated with chemical fertilizers. Obviously, the over-consumption of smoked and manufactured meat products, with all of their carcinogenic additives, should also be minimized.

We must, of course, respect people who refrain from a meat diet because they do not wish to support the inhumane treatment of animals in modern factory farms, because this is a gross violation of the cosmic laws, and it is understandable if people are 'put off' from eating meat especially if they have witnessed these practices first-hand. However, a meat diet should not be abandoned – rather, these practises should be changed. By choosing quality over quantity, even if it costs a little more (although, in fact, less is needed), consumers can do their part in forcing change.

A close bond of body and spirit (underpinned by a balanced blood composition) sharpens the mind and intuitive sensing. In tense situations the blood radiation is strengthened by hormonal metabolism, mainly by adrenalin (epinephrine), to mobilize the body. Then the muscles are tensed and attention is focused in a realistic appreciation of the challenge at hand. If this increased alertness and appropriate reactions enable the individual to overcome the threat, it is said quite rightly that he has shown great 'presence of mind'.

Lines of lively resistance

In sleep, the union of body and spirit loosens. The body does not bind the soul as firmly as in the waking state. In falling asleep, the vital functions slow down, the heart beats more slowly, the body relaxes, and its temperature drops. Slowed brain wave patterns correspond to the stages of falling asleep, the slowest waves coming in the state of deepest sleep. Activity of the brain and other organs is weak during sleep, when blood radiation also lessens and the body–spirit union loosens. The sleeper feels that he is falling – hence the expression 'falling asleep'. The impression of falling takes the form of a 'jerk' as the sleeper feels the loosening between body and soul. There is a sense of free-fall or falling in space that the body tries to resist, as if attempting to reassert itself. In the dreaming phase of sleep, manifested mostly by rapid eye movements, or REM, the bond between the body and spirit is at its furthest stretch. This stage is reached more quickly in a room with a low temperature, so that body temperature drops quickly, and facilitates the loosening of the bond.

The human embryo, in its earliest stages of development, is unable to provide a firm bridge for the spirit to incarnate in the womb. Its blood is very limited in volume, and it does not exist in its final composition. What we call 'human life' requires a union between body and spirit. A child inherits a body, but it conditions its own blood radiation. The developing child comes along in its own particular personality, which is little affected by its parents. How, then, is the body–spirit bond formed in the human? Tradition tells us that at some point, possibly at conception, a soul is breathed into the being. This explanation serves to distinguish people from beasts, and flatters the superior creature by giving him something extra, a spiritual nature that animals lack. The animal too, has a soul, which incarnates and then departs again at death, but the soul of the animal is a totally different species to the spiritual core that animates humans. It has no free will in the human sense and is therefore not personally responsible for its actions in the way that humans are. Therefore it gains no benefit from undergoing suffering, and – quite rightly – suffering animals are 'put down'. Attempts to carry this over to humans, as the advocates of euthanasia and their supporters propose, demonstrates a lack of understanding of our spiritual nature with its accompanying free will and the consequences of its use or misuse. Implementation of euthanasia should be opposed at all levels.

In a materialistic world, expositions on the nature and activity of the spirit are always likely to seem unconvincing or far-fetched, since they must by their nature be dealing with affairs that are immaterial, that is, out of this world, and can therefore only be fathomed by our intuition. If we accept that the spiritual is the essential part that endures in the ebb and flow of our existence, the only ultimate form, then, using the tools of science and reason and the metaphors of art ('metaphor' etymologically meaning 'bridge above'), we may be able to approach the core of the mystery of our transient wandering on this earth. Hence we speak of radiation that can be detected, for example, or of the blood in mental or organic disease states, or a person's aura, and we speak also of the 'bridge' that can neither be seen nor measured. In preparation for its incarnation, the spirit will use a temporary bridge of the mother's blood in order to bond with the embryo's growing body. The mother's blood does not stream through the embryo as the mother and child have separate circulations that are isolated from each other. Instead, the mother's blood streams through the placenta, which provides an exchange surface between the mother and the fetus, furnishing it nutrients and providing all its nourishment.

The mother's blood radiation permits the spirit to approach the body growing in her womb. The attraction commenced from beyond the earthly even before procreation occurred. A large number of discarnate souls in the immediate beyond stand in anticipation of becoming incarnate again on earth. This might be for reasons such as the redemption of karma or reciprocal threads of fate. Initially, this union with the mother-to-be is loose, still a long way from the incarnation, and the embryo is yet to attain a form that can accept the spirit. However, the spirit already influences the development of the body growing in the womb.

Incarnation proceeds when the fetus has developed sufficiently to offer the spirit a vessel together with the necessary bridge of blood radiation. At its earliest stage, the embryo is but a bunch of dividing cells, and before acquiring human form it will pass through various stages of animal evolution. Initially it will resemble aquatic creatures like molluscs, fish and creatures with gills of some description. Loosely speaking, it then takes on features found in amphibians, then those of land animals, progressing up the evolutionary ladder to mammals and primates, and finally humans. At the end of the first month of gestation the organs take on their gross form, and over the next two to three months, up to the middle of pregnancy, they will perfect themselves in detail. Up to this stage, fetal development is essentially animal in nature. From the fourth month onwards, the tissues and organs of the fetus, even their innermost structures and even if many are in primitive form, are almost identical to those of the newborn.

Blood vessels form early and are derived from channels formed within clusters of primitive cells in the middle layer of the early embryo (mesoderm). These channels extend and fuse together to give rise to an extensive network of vessels; after the second gestational week primitive plasma and blood cells develop from the lining cells of these vessels. Blood, although present in the blood vessels, does not circulate at this point. The first heartbeat does not occur until the fourth week, and the heart only starts beating effectively at the end of the fourth week, thus initiating the flow of blood. This blood does not have the same composition as that of an adult. The red blood cells are formed from clumps of cells within the primitive blood channels, and these first cells possess a nucleus. By the fourth week of gestation, red blood cells with no nucleus are being formed. From the fifth week onwards, blood formation takes place mainly in the liver, and from the third month it occurs in the spleen, bone marrow and lymph tissue. By the fourth month nearly all the organs involved in blood formation are active in the fetus. The blood is sufficiently developed to form a radiation bridge for the spirit. The body is ready for incarnation. At this stage, what we are accustomed to calling the soul enters the body. That first 'shock' or 'twitch' is felt by the mother, with feelings of joyous elation or despair, depending on the character of the soul that has entered and her attitude towards the impending birth.

Thus the spirit neither incarnates at conception nor at birth, but *midway* in the pregnancy. At fertilization, only the *possibility* of incarnation exists.

Falls the shadow

Not all souls are benign, and we now have to consider the ways in which so-called evil spirits may work on a body with an altered blood radiation. When we speak of 'evil spirits' we are not necessarily assenting to the presence of black magic and the powers of witch doctors; rather, we refer to the fact that through the ages people have been known to be 'possessed', in various ways. There is a logical explanation for this phenomenon. A weak blood radiation renders the body vulnerable to influences from discarnate souls seeking to dislodge the occupant. Such an 'intrusion' can account for much bizarre behaviour, menace and terror as experienced in affected households.

An intruder wrestles for total or partial control of an earthly body by means of unnoticed influences outside awareness. Young people undergoing changes in blood radiation and temperament during the years of puberty are often affected by such an intrusion, especially if they lack high ideals and are easily swayed. The intrusion can also affect an adult who lacks spiritual vitality because of inner spiritual (but not necessarily intellectual) indolence. The body is protected by a secure inner attitude. In certain circumstances, strengthening or changing the blood radiation, for example by taking certain medicinal preparations or special diets or nutrients, is useful in defence. However, only a physician who acknowledges the existence of the metaphysical and takes an integrated approach to healing will find the correct combination. Such persons are usually, but not exclusively, found in the ranks of those practising natural medicine rather than orthodox medicine.

Alien seizure in which the occupant is dislodged can be brief or long in duration. The encroachment stirs an inner conflict, the first signs frequently showing in the brain. This internal conflict may manifest as attention and concentration difficulties, a lack of logic, incoherent speech, absent mindedness and withdrawal from the world. The person who is affected may become unnaturally passive or over-agitated, sometimes alternating between the two extremes, thus showing classic bipolar manic–depression. Lack of consistency is the hallmark, together with unforeseeable reactions, bizarre and contradictory behaviour and weird notions. In extreme forms of possession the victim may have to be sectioned (admitted involuntarily to a medical unit for the evaluation and treatment of a psychiatric disorder) for their own and others' protection.

The abnormal behaviour can be persistent or occasional. From the outside, the individual appears irrational, and is deemed insane; this insanity is borne of conflict for possession of his body and brain. Several souls may vie for possession of the brain, so the possessed individual may at one moment be gentle and considerate, only to change in a flash to a violent and impulsive personality, then become fine and cultured again, to be followed by an outburst of vulgarity and cruelty. The possessed person can, by turns, give the impression of being well informed and then be utterly ignorant. Up to a dozen different personalities have been identified in one sufferer.

Various theories have been put forward by medical practitioners who are unwilling, or unable, to acknowledge the existence of the soul or of the 'beyond' and its inhabitants, in order to explain these phenomena. However, they are all way off the mark, losing themselves in high-sounding phrases and technical jargon. The real cause is not understood by practitioners of orthodox medicine, so very little can be done to help those who are afflicted. Such patients (one becomes a 'patient' only when admitted to hospital) are commonly described as incurably insane, although some are only temporarily possessed and may benefit from a change in their blood composition to correct the radiation. Such treatment is futile in cases of insanity that is related to an organic brain condition.

Many so-called ‘mentally backward’ children may not in fact have a defective brain; rather they may have an unsuited or impaired blood radiation. In reality there are no sick *spirits*, just abnormal conditions arising from dulled and coarsened coverings around the spirit. This abnormality is rooted in the spirit’s inactivity and is the certain basis of evil, which a lazy spirit is unable to ward off. Unless the spirit is forcibly hindered by organic brain disease, its inaction will account for much mental dysfunction. In a few instances where disease, diet, habit may have adversely affected the blood composition, changing the blood radiation will alleviate things.

To sum up, human blood serves mainly to provide a bridge for the activity of the spirit in this world. A human being is not just a physical body of flesh and bone, but is at its core an immaterial spirit that has incarnated in a body. The spirit is enveloped in different, but closely linked, fine coverings of the consistency of one plane to the next acquired during the descent from its origin. On earth it takes up the physical body. The intimate connection between the spirit and its coverings is responsible for our state of consciousness. Our minds process prodigious amounts of information. The spirit operates through its coverings in the earthly. Its will passes to the brain, which then sets the body in motion. In reverse, sensory perception is processed toward the spirit. The union of spirit and body through radiation can be strengthened, changed or distorted by behaviour, attitude and outlook. For example, spiritual indolence is marked by rigidity and narrow-mindedness. These people are often characterized as requiring a clear set of rules and guidelines as well as a routine in order to function. Anything out of the ordinary throws them off-balance. They lack common sense as well as ‘presence of mind’, both of which are spiritual rather than intellectual qualities. Mental and physical ailments may be tended by diet and medicines from nature's vast pharmacy. Then an inward disposition for the good sends out a warm glow, which through the blood radiation is able to invigorate the body. The will of the sufferer, set spiritually aglow, assists the cure, restoring the equilibrium of mind and body.

The immune system does much to defend and protect the body against disease, aided by hygiene and sanitation, improved nutrition, vaccine technology and nature's medicines. Excessive reliance on antibiotics, however, has given rise to drug-resistant strains of bugs that are even more difficult to control. The immune system that relies too much on outside help in the form of processed drugs may become sluggish and less effective. No drug has yet managed to completely eradicate a disease. The spirit has a temporary dwelling in the physical body and its manifest will depends on the quality of blood radiation available to it. The human blood can form only when the spirit enters the body. When the spirit departs, the blood ceases to exist as such.

Island in ocean of ignorance

The first humans inherited their physical body from the hominid apes that roamed the earth millions of years ago. A fully developed brain capable of abstract thought and the capacity to question and wonder was required for the human being to develop spiritually. The human brain is different in many ways from the brain of the anthropoid ape, for example, it is three times larger than that of the chimpanzee. Up to four million years ago, our hominid ancestors had brains that were about a third to a quarter the size of the modern human brain. They made and used stone tools and had an awareness of time, which is a uniquely human characteristic. The two brain parts – the cerebral cortex and cerebellum – grew in unison in early humans, as the connections between the neurons grew increasingly more complex, so that early humans had a more evenly proportioned cortex and cerebellum than we have today, with our relatively enlarged frontal brain and stunted cerebellum. Thousands of years of over-cultivation of the intellect, with intuitive recognition sidelined, led to this over-enlarged frontal brain, to the detriment of the 'atrophied' cerebellum, the much neglected spiritually receptive brain. This smothering of the intuitive perception, which should actually grow in its immediate insight to guide decisions in life while the intellect executes with observation and reason, amounts to incarceration of our spirit and constricts our consciousness to the transience of material space and time. The cerebellum of pre-modern humans might have afforded good vision, for example, especially before the development of cultural skills like the mastery of fire. Large-brained modern humans have a highly developed ability to plan and organize, but states of altered consciousness, as in dreams, are mandated by the cerebellum.

Deeply moving impressions become personal possessions, or as we say, they 'take possession of us' and become part of us. We absorb fragments that feed our souls, as in artworks, ideas and impressions. Intellectual knowledge, on the other hand, is mostly borrowed property, recycled thought-forms deriving from the experience of others. It belongs to the material world rather than the world of spirit and it will perish with the body. If written down, this knowledge can be passed on to future generations to build upon so that it is not necessary for each succeeding generation to re-invent the wheel. Strictly speaking though, all intellectual pursuits, regardless of their field, have the sole purpose of alleviating life on earth, of freeing the human being from performing menial tasks or enabling him to 'get about' more quickly. They should never be considered as the *sole* purpose, or even as the *main* purpose, of human existence.

The intellect, dominant in our modern condition, restricts, dissects and analyzes perception in the frontal brain, without interacting with the intuitional brain to enable the greater cognitive capacity of spiritual recognition to come into play when making decisions. Early humans, guided by intuition, thought more in pictures. As intuition strengthened, their instincts were subordinated. Thousands of years of excessive straining by the intellect has resulted in our enlarged frontal brain and stunted cerebellum. In fact, eighty-five percent of our brain consists of the cerebral cortex alone, a distortion that heavily impairs our cognitive capacity.

This brain distortion has, over the ages, become hereditary. The child is burdened by the inheritance of a large brain – the mother with a long labour and painful childbirth. By the end of the fourth week of gestation there is already a sprawling development of the frontal brain overlaying the other parts. Unlike other creatures, the human brain mainly develops after birth. But the frontal brain of the embryo already contains one hundred billion neurons. Within two years of birth these become a thousand trillion connections in a gigantic network of modulatory systems. From the outset the child is predisposed to domination by the intellect. Therein lies the basis of inherited sin, in which the burdensome subordination to the intellect orients humans away from the Will of God, which rests in the cosmic laws. By misuse of our free will, we changed ourselves and our species in a created order of the cosmos permeated by unchanging laws. Thereafter we must incarnate in the world ethically branded by acts of our will. The culpability for the burden of an over-cultivated intellect derives from our own individual actions.

For millions of years we have wandered through this world and the beyond. At the beginning we descend, as unconscious spirit sparks into material substance. There, the unconscious spirit germ is awakened by experience to conscious activity, and should adjust voluntarily to the cosmic laws in the upbuilding in Creation. To insist on proof of this unthinkable process is to invalidate the intuitive powers that are vouchsafed from the spiritual realm. Poets, philosophers and seers for centuries past have attested to these very real powers.

The earth is the furthest outpost on our migratory wandering through the material planes and it is the turning point for ascent to the spiritual home. The long journey of our development in different planes in order to become completely aware of the laws in their cosmic manifestations is incomparable with earthly time spans. We cannot mature in spirit in one earth-life. There are many rungs on the ladder.

To continue this metaphor of ascent, we need two feet in order to climb a ladder and each foot must take its appropriate share of the weight burden, so we need to keep our balance between intuitive perception and rational thinking of the intellect, and the complementarity between positive and negative. Looked at another way, with a different metaphor, two poles are needed for electrical current to flow. We are binary creatures of male and female identity, split species incomplete in ourselves, striving towards each other, seeking completion.

Beings darkly wise, and rudely great

We descend as spirit into the world like a seed, not self-created but endowed with abilities to attain a self-conscious personality. Our spirit germ, bombarded by powerful vibrations of the changes of evolution and dissolution in the material world, awakens sensitive volition, analogous to how a seed planted in the earth is stirred to germinate by the elements. The seed must help itself if it is to germinate. Our spirit germ exerts its ability of free will in favouring certain vibrations, and is subject to the reciprocal effects of the choices it makes. In becoming conscious we incline to a certain nature of activity. In this way, we decide early in our wandering on masculine or feminine activity, and our choice will be permanent.

We start out as spirit germs without any gender but with the urge to become conscious. As with a grain of wheat that bears the potential to become a ripe ear of corn, so it is with the spirit germ unfolding to become conscious. Male or female gender develops from the nature of activity; male corresponds to the positive pole of spiritual activity, and female to the complementary negative pole. The word 'positive' is used in the sense of vibrant and earthy, and 'negative' in terms of delicately receptive and uplifting. Our gender and sexuality ought therefore to be in accord with the nature of our inner activity. Womanhood is endowed with a capacity to express the stronger intuitive recognition by virtue of stronger upward connections with the finer currents from our spiritual origin. Feminine activity has the capacity to refine the more earthy masculine activity. Women can, with the delicacy of intuitive perception, take the lead in the ennoblement of their surroundings.

We should pause here a moment to consider the nature of the procreative urge among humans, a force that has enormous potential for rising to sublime heights or sinking into the most abysmal depths. Like us, the animal is subject to the sexual instinct, but unlike us, beasts experience no shame in indulging this instinct. This fact shows that our main vocation is not just to reproduce. Excessive indulgence of the instincts is a hindrance to spiritual unfolding. So, too, our passions can infiltrate our intuitive insights, dimming them and leading to erroneous judgements, more so because the self may be viewed as derived from an assemblage of sensations and passions.

Life on earth, being all motion and development, manifests by deviations and differences. Spirit germs, on the other hand, starting out unconscious and without gender, are impelled by their urge for development to wander in the lower parts of Creation. Thus a neutral state finds fulfilment in positive and negative forms, and dynamic processes are set in motion, one of which is the sexual complementarity evident in nature.

A note on homosexuality and bisexuality may not be out of place here. Such deviations from the norms and reciprocities of human development have received much attention.

Marching under the banner of liberalism and tolerance for minorities, departures of this kind have in our time acquired almost cult status. In the homosexual, the inner and outer self are a mismatch, so this person is a stranger to himself or herself, because a man can never truly become a woman, nor a woman a man. A homosexual man strains his inner nature by weaving an effeminate mantle around his masculine core, for his spirit will always be male. So, too, will homosexual women incline to or emulate masculine activity, against their feminine nature. The spirit is hampered, and the distortions result in gender conflicts between self-identity and the body. The body-swaps of gender reassignment procedures attempt to adjust the body form to the innate male or female self-identity. Alternatively, the readjustment may come with reincarnation.

Transsexuals, that is to say people who feel they are of the opposite gender to their physical sex, are usually souls who were probably on the point of recognizing their error before incarnation into the earth-life, but the volition was perhaps weak. Homosexuals and bisexuals may pass through life more or less 'well' overtly, but psychically distorted personalities tend to be drawn subconsciously to each other in a relationship. The two are united by a mutual understanding. For instance, a woman who harbours a distorted male soul will often be drawn to men who harbour a distorted female soul, and vice versa. Such compensating weaknesses may well work to bring about a strengthened natural state of the soul, so that the mistake may not be repeated in another incarnation.

Relationships between complementary spirits are naturally bonding. The urge for harmonious union of complementary souls for mutual development penetrates to earthly bonding. The procreative instinct is of secondary importance in the bonding of love, companionship and true affection that absorbs the whole being and springs from the spiritual urge for completion. The complementary interaction can also involve two people on opposite sides of the world who are strangers –

and between spirits in this world and the beyond. Thus, while physical relations between the sexes serve the procreative imperative, the primary purpose is the exchange of what we might call ethereal forces in the material release of mutual love.

Homosexuality is neither natural nor inevitable in the course of our development. A homosexual person, feeling trapped inside the wrong body, indicates some distortion in the coverings around the unchanged core of spirit. Today, moral censure is muted by tolerance and compassion. The negative side to all this is that by portraying these deviations as perfectly *normal*, even 'trendy', those who are affected never come to reflect upon the nature of their condition. The fact that a great many homosexuals have been instrumental in furthering the advancement of the human spirit – in the fields of art, science and religion, for instance – may suggest that the tumult of sexuality and gender impels towards an inner awakening, setting the individual soul back on the path towards emotional and bodily alignment.

Embracing the gift of life

Our birth into the world initiates just one segment of our existence – it is not the very beginning of it. The child inherits nothing of the personality of the parents, but is an independent human spirit in the course of its own development. The newborn is in reality an old soul, a spiritual personality in its own right. It inherits not the least speck of the parents' spirits. It receives only the physical body, the anchorage it needs for its own spirit's activity in this world. The child can just as little credit his parents for any inner abilities as reproach them for any inadequacies. The newborn is not a tabula rasa, neither is it innocent or without threads of fate. Exactly as its parents did, the newborn carries weaknesses and any developed abilities, therefore its own karma, both good and bad.

Human qualities are expressions of the soul, acquired by experience in the course of our wanderings and they remain the personal property of the soul. Pure or otherwise, our inner qualities are personal and without hereditary transmission. Clustering of abilities in families, music in the Bach family for instance, is a matter of the attraction of homogeneous souls. We tend to associate with people who have attitudes and behaviours like our own, and are most attracted to those whose inner qualities are similar to our own.

Our existence covers several incarnations. Just as, at death, and if not held earth-bound by our threads of fate , we cross to the ethereal plane, the so-called beyond, so at birth we come into this world as independent spirits. Being on earth today means that we have dispositions and propensities to dispense with before we can attain the higher steps of development towards the spiritual sphere of our origin. Children in their early years are more sensitive to animistic than spiritual vibrations in their intuitive perception and learning. Although enlivened by spirit, animistic perceptions dominate their thinking. Their immature physical bodies cannot yet enable the spirit core to step forth actively, in its full estate, so to speak. The child, therefore, has little more responsibility than the most highly developed animal. Perception in the very young is largely animistic. Thus, rather as an animal will forever seek out a certain place or tree, a child is in direct touch with nature and will find magic in field, pond or wood. He will recognize and relate to certain animistic helpers active in nature, such as elves or gnomes, often not seen, and withdraw in fear from others, such as nixes (water sprites) or trolls and so on. Thus parents and educators and especially those responsible for town planning must ensure that children have access to these places, which can be a real challenge today as parklands and open spaces give way to housing developments in our burgeoning cities. Sophisticated toys and computer games provide a poor substitute for these open spaces.

Such an early relationship with the wonder of nature provides the right impulses for the mental development of children and prepares them to adjust to adulthood. After all, childhood perception also encapsulates the collective consciousness of our ancestral experience. Thus spiritual man is rooted on earth in animistic realities, as evidenced in folktales. But this foundation has in many cultures been diminished to myth and superstition in the name of civilization, especially driven by secular and ecclesiastical authority and missionary zeal. Modern humans are thereby tragically disembodied at an early stage from their animistic rootedness. All too often this independent development of a child's personality is invaded by adult minds with no heart and time for the child's ability to engage the world, and the spirit is overshadowed by cerebral perplexities. As Wordsworth said, 'Shades of the prison house begin to close upon the growing boy'. Adults deprive youths of much that used to belong to the teenage years. Instead, children and early adolescents are left to grapple with issues and problems that undoubtedly belong to the adult world.

Something is clearly lost in the transition to adolescence, but something is also gained. Puberty sets in with its storm of hormones raving through the body; the body changes rapidly and the brain is enormously remodelled as many thousands of neuronal connections disappear every second. With physical and mental development comes an awareness of the self and its identity, together with intimations of spiritual power and personal destiny. Physical development lowers the metaphorical drawbridge that allows the spirit to interact with the world. Adolescence is equivalent to being born again. At this stage, youths also develop recognition of their own sexuality and its effect on others and on the environment at large.

The stirring of the spirit in our adolescence is manifest in the yearning for ideals, an exacting commitment to authenticity of the individual self amid the norm of reciprocity in the social environment. Youthful idealism carves out a path to awakening in the psyche that should be travelled on solid footing towards the maturity of adulthood. Unlike adults, who rely predominantly on their intellect, adolescents use more of the cerebellum to process information; they are, or should be, more intuitive, less cerebral. However, if the intellect is schooled obsessively from too early an age, as is often the case today, this yearning may be subsequently stifled or may find its expression in an aggressive attitude, through such things as sport where winning is all that matters and the struggle for self-improvement is not against oneself but an all-out struggle against others, or a general disregard for and non-acceptance of the authority of adults.

Therefore the awakening, which should be a force for good, is often short-lived. The zeal fizzles out as the spirit lapses into indolence in the wake of the surge of the intellect together with constant seeking for thrills, stirred especially by the massive brain changes. And as much as the restructuring of the brain may afford the youth greater mental self-sufficiency in order to make increasing complex decisions and analyses and to lessen the influence of adults when faced with life choices, we find that this coming of age results not in positive upward motivation, but in moodiness, melancholy, brooding and self-indulgent rebelliousness. In extreme cases, the youth is characterized by temperamental instability. Of course, a healthy dose of melancholy, if kept within bounds, is to be welcomed, especially as the melancholic temperament at this time gradually replaces the sanguine temperament of the childhood years. Conflicts mount up with the youth's parents, particularly because the adolescent insists on being treated as an adult and not as a child – even more so when perceived rights are concerned. Once responsibilities are mentioned, however, many of them are happy to become 'kids' again. There is discontent, rage and wrath, with edgy dialogue. The youth revolts against accepted modes of thought and norms of behaviour and breaks with parental conventions in a restless struggle of self-identity. Demarcating the world of adults becomes a path of trial and tribulation. On the street corner and at the rear of the classroom we find teenage angst, edgy youths cast adrift in the menacing uncertainty of early adulthood. The surge of sexual energy is like physical fitness, at its peak in the teenage years. But this life-affirming premium, by which the emergent spirit should take control of the mature body, is dissipated in excessive thrill-seeking. The spirit, now indolent, gets choked off and becomes entangled in the physical. It languishes in the moral fluidity of material aspirations that are wholly dissatisfying.

Unfortunately this so-called youth culture, proudly wearing its badges of alienation with a complete lack of self-justification, is patronized by the media for its own commercial ends. Magazines, record companies, the press and the television and film industry have over the years pandered more and more to the gross and violent appetites of the young. We know that a relationship with shamelessness and vulgarity has come to overshadow the media, yet even to mention this obvious fact or cast doubts on its virtue is taboo. Like the spectators at the parade of a naked emperor, we are constrained to applaud his new clothes. The intuition of shame, which restrains capitulation to the purely physical, is subordinated to the influence of thoughts, even outside one's awareness, and social trends in the guise of progress and sophisticated modernity. The instrument of the physical body for the human spirit is jeopardized and the noble qualities of the psyche are mocked and deliberately suppressed.

To respect the body is to respect the soul. Conversely, abuse of the body, including alcohol and drug abuse, wearing outrageous outfits and make-up, undergoing piercings and tattoos and self-mutilation, as well as excessive indulgence in sex, is abuse of the soul. It is true that government initiatives, in schools, surgeries and elsewhere, have attempted to draw attention to the more extreme forms of bodily abuse, but these programmes all too often merely tinker with the problem. People are told how to have better sex, safer sex, more sex, all of which encourages indulgence, but none of which aids self-understanding. Sex education is in general aimed at preparing young people to make informed personal decisions in their attitudes and behaviour to sex and relationships. Yet as a school subject it puts the emphasis on the body. It isolates an area of human activity from its native grounds, for sex in the human context is not simply a matter of copulation. Nature provides the right education in this respect. Through observation of mating among animals, children will of their own accord draw conclusions about human beings, aided by questions that are answered truthfully and tactfully by a responsible adult. For children to observe humans mating is entirely another matter, piquing sexual curiosity prematurely, although it is probably less harmful than the rash of hints and sniggers embedded in sundry comedy shows.

As mentioned, there is a strong bond of animistic affinity between the child, whose spirit is yet to unfold, and animals, whose animistic core is embraced in a group soul. But animals do not have a long trying prelude to adulthood; they are kittens or pups, and then they are mature. Things are very different with us. Child upbringing prepares the soil, as it were, for the awakening blossom of adolescence, when inner vibrations and outer influences impinge on the previously sheltered spirit. Adolescence is a crucial period in human development and it should not be left to look after itself. Above all else, it should not be exploited for the vulgar and material ends of corrupted adults.

Stable of neglect

The term ‘struggle for life’ evokes a picture of combatants scrambling for finite resources of nature. There is within the term an implication of savage injustice, that in the end the strongest will survive and the weakest perish, that 'might is right' and there is no such thing as natural justice. Rather, the imperative of movement permeates the entire creative order of the cosmos. The spur of competition forces the creature to bestir itself, to be alert, and this, positively in its turn, guards against psychic stagnation.

We move or we perish. This great impulse eschews the deadly embrace of sloth; it induces in creatures, big and small, the imperative of constant exertion and alertness in order to preserve life. It urges us to shake off fetters of heredity or environment. But the freedom for one to develop may well incur adversity and setback for another. This presupposes moral reflection and a sense of fair play. Deep down we all know that equal opportunities should be available to all. This does not mean, however, that differences in wealth and status should not exist, because we all weave our own fate by our actions – even if they occurred in a former life. What is meant here is that nobody should be deprived of the essentials of life such as food, shelter, medical assistance, basic education or training and the opportunity (or compulsion!) to do some useful work and be fairly compensated for it.

However, on the face of it, this innate awareness seems at odds with nature's essential competitive urge, the drive to survive and prosper. And since absence of competition is detrimental to spiritual mobility and development, why should we heed the voice of conscience? For surely indolence will thwart the state of general well-being and confidence. A culture that is averse to personal enterprise and progress will be marginalized.

What we know should *be* and what we see that *is*, as so often in this 'fallen world', seem to be at odds with each other. Unrestrained competition is at odds with humanity and decency; it ignores the voluntary balance between giving and taking in the fellowship of love. Unscrupulous behaviour will ultimately get its comeuppance. Wherever we find one-sided or ignoble competition, where one person is pitted against another, aggression erupts, with war and suffering. The savagery apparent in the wild kingdom is no model for humanity, for we have the compulsion to absorb finer radiations, which we not only need for our nobility of character, but are also obliged to mediate to lower creatures in the ennoblement of our surroundings.

This higher perception is often lost to us. Not only in our dealings with our own kind, but also in our interactions with the environment and the animal world. Cruelty, sometimes heedless, sometimes deliberate, seems to be our prerogative. For instance, while animals kill for food, we often kill for fun. We will heedlessly wipe out species, even feeling triumphant in wiping out our own. Animal predators make a quick kill of their prey; but only we enjoy inflicting a slow death. Animals operate within certain rules which advance the interest of the species; humans are not always so consistent. The fighting instinct is a spur to keep the creature inwardly awake and vigilant. It preserves the inner movement needed for survival and development of the species. One obvious example of this is the strong predatory pressure that facilitated development of the antelope into a nimble sprinter.

We are by virtue of our origin and our faculties unavoidably dominant over all other species, but we must fully assume the responsibility we bear to guide and to lead the material world to further ennoblement. Our living core is spirit, which through the vessel of the body experiences in the world in order to attain self-consciousness. Spirit runs through all of nature, attaining self-consciousness in us. By being in touch with our spirit through being more intuitively perceptive, we can be said to be bearing a torch for the Creator.

All passion spent

The cosmic laws work to further all possibilities of natural development. They press toward the ideal and are at work not only on our small planet but uniformly in the whole cosmos as manifestation of the primordial, never-changing laws resting in Creation. Untold marvels are revealed by science, seemingly fixed laws of physics and chemistry have become fluid, and our vocabulary is stretched to keep up with these discoveries. Thus we have 'curved space', 'time warps' and such notions as 'before the birth of time'. It still remains to be seen if these expressions, which attempt to denote things we cannot grasp with the intellect unaided (and are therefore paradoxical), will stand the test of time. For as the Apostle Paul once said 'For we know in part ... but when that which is perfect is come, then that which is in part shall be done away' (1 Cor.13:10). 'That which is perfect' can only refer to new knowledge from a vantage-point far above our sphere of origin mediated to humankind. It will give us insights into the workings of Creation that we would otherwise never attain – for our science can only know such things in part.

Unlike billions of the teeming organisms on our planet, we at least know that we are alive. But what is life? What are we to make of it? Is it merely an energy of motion in a soul-less process of animal electricity? Is it a reality of itself? Or is it a nebulous abstraction from the highly complex states of organic matter, the improbable outcome of an immensely long evolutionary process of Darwinian self-organization? These questions have been asked by people for centuries, and modern science is no nearer to a final answer than were the ancient civilizations of the Greeks, the Egyptians and the Chinese.

In all the great cultures of history there has been a dedicated confrontation with the basic nature of the world we inhabit and how it developed. Today, scientific arguments in favour of an immeasurable expanse of the known universe sit shoulder-to-shoulder with received ideas of a 'heaven', of God and angels above this world, along with novel and even esoteric mixes of alternate realities existing laterally with the earth we inhabit. For example, some expect Jesus to return as the Son of God from the summits in blazing glory as a spaceship captain who will save humankind with advanced aircraft from the fallen state. We humans know but little of the structure of Creation. Some ancient fundamental conceptions of reality are today regarded as myths and legends that cannot be taken seriously, others are still considered to be religious truths, such as the idea that the created order of the cosmos came by a command: 'Let there be light!'.

Abd-ru-shin states in *The Grail Message* that the expanse of the visible universe, which modern science studies so intensively, is but a small part of God's Creation! Such a sobering thought leaves but little room for our significance as the highest life form on this earth and how we have traditionally seen ourselves as existing at the centre of Creation. All spiritual conceptions that can be taken seriously consign the hustle and bustle of our physical world, even the orbiting billions of suns and planets that swarm and cluster in galaxies too numerous to count, to the status of a sedimentary formation in the all-encompassing motion of Creation.

On the vexed scientific question of whether there was already something 'before' the creative act that originated the observable universe, *In the Light of Truth: The Grail Message* affirms that there was always 'the Light', which is at the same time God, the infinite One, Whose unavoidable primordial radiation, sublime and independent, is also eternal, thus without beginning and without end. God alone is Light and Life. With this 'radiation', however, it is not Creation as we know it that is described, but only God's immediate vicinity of the everlasting Divine Realm. The work *The Grail Message*, describes this as the divine substantiate – as against the indescribable Divine *Un*substantiate, God. Here, in the midst of the purest, in the immediate vicinity of Life, exist beings who are endowed with the highest consciousness, including those called 'archangels' who originate from the direct radiation of Unsubstantiate Life. The expressions 'substantiate' and 'unsubstantiate' are certainly not in common usage but they were chosen by the translators of the text as the most appropriate words in the English language to represent the original German expressions used by the author. It is only when taken in the context of the lectures contained therein that there gradually arises within us a faint conception of what is meant by these expressions.

Life, which to us manifests as eternity and infinity, is the omnipotent Creator alone, that indescribable being that was once upon a time too awesome, too holy to name – since when we have the ninety-nine names used for Allah, or Yahweh, Jehovah or what nowadays we have come to call God. What we know of 'life', Abd-ru-shin explains, is but impelled movement, that which can be seen and recorded by mortal eyes. In reality this movement, this potential for change, has its origin in the power of life pulsing through Creation as the driving force for its coming into being and further development. Whatever images we may raise to worship or idolize, from Aztec mask to totem, or golden Buddha to the Crucifix, human spirits can never successfully conceptualize the Creator of all things. Even from our highest perspective, God remains a disembodied presence beyond intuition, the unknowable. A conception of God as Nothingness has also come about in part because the ludicrous and childish notion propagated by the churches of the 'Old Man upstairs who runs things' can no longer be upheld in this day and age.

For God can neither be seen nor known personally by human spirits. God can be perceived only through the radiation that emanates from Him and manifests in the entire cosmic order, to which all creatures owe their existence. Thus Walt Whitman says he finds letters from God signed and discarded in city streets. He might equally have found the language of God written in every atom, every molecule, every light wave, impulse or passing thought. Life, God, cannot but radiate, and His Power manifests immediately as the eternal Realm of the Divine, after which everything else in existence likewise, although in descending gradations of 'aliveness', originates in the pressure of His further radiation. This radiation is not His Being, however, but only the consequence of It. The highest possible recognition of God for the human spirit – and it is truly the spirit and not the intellect that can attain this recognition – lies in the recognition of God's Divine Will, which manifests in the unalterable laws resting in Creation, the final ramifications of which are the natural laws that we come to recognize in our perception of the physical world. Thus the existence of God does not originate in our thought-life, which itself is at best only a consequence of our deepest perceptions, but in the unsubstantiate reality beyond the speed of thought and the capability of our conception.

In the beginning, let us say, for of course life has no beginning that humans can ever pinpoint, the primordial stream brought light and therefore life to the previously formless void beyond the divine sphere. Upon the Creator's act of will, Creation came into being from transition of the radiation outside the divine sphere. The radiation cooled and diminished in pressure over inconceivable expanses, encompassing worlds of distances without horizons, and dividing into planes or spheres with seeds of living forms. Planes formed with gradations as Creation expanded in the pressure of the radiation of God. The more distant the plane of Creation was from the summit the more the transformed radiation cooled, with a corresponding deceleration of events. The plane's consistency corresponds to its inherent mobility, which, in turn, determines the conceptions of time and space. Each plane is adjusted to the attraction from the summit of Creation. The infinity of God is not anchored in the finite so as to imply any interdependence and intimacy with Creation. Rather, all that exists finds its origin in the power of God's radiation. All Creation is pressed in motion in the radiation of God.

Intimations of infinity

In the divine ring of God's immediate radiation entities arise that can bear the enormous Light pressure of the proximity of God, although this proximity can only be taken allegorically because even there the distance from God is so great as to be virtually inconceivable to us. All else that can possibly exist, but which cannot bear the divine pressure in the proximity of God, is held in the residual radiation. We could compare it to the aura of an extremely strong-willed personality whose pressure on others around him leaves weak human beings with hardly room to breathe, and who need distance in order to be able freely to become aware.

God gave release for the residual radiation in time and space. Creation came into being from the transformed radiation outside the divine. First there were the primordial planes at the summit, embracing the purest spiritual extraction; then followed the lower spiritual species, its furthest planes giving origin to our developed species, which issues from the residue of the transformed radiation that is unable to develop to consciousness within the bounds of the spiritual realms. Other species could only develop at even greater remoteness from the primordial summit of Creation.

Language is inadequate to capture the vitality inherent in these planes of Creation. It is just as impossible for an artist to depict the living torments in the dark regions as the rapture which life holds in the luminous realms. So, for example, the artistic gesture towards the image of ridiculous plump, harp-playing angels that today evoke boring ideas of monotonous 'heavenly joys'. The primordial spiritual beings at the summit of Creation are the ones made in the image of God. Their volition gives form to realms below that include the cradle of the human species. According to *The Grail Message*, we human beings originate as spirit 'germs' that require a great remoteness of tremendous cooling from the Light pressure of the proximity of God in order to develop the urge to consciousness. Following this urge we journey from the spiritual realm to be immersed in lower-lying and less mobile material planes, in order to awaken and mature our spirits. Awakening comes about through the friction generated between ourselves and our coarser surroundings, forcing us out of our inertia just as our physical body could not move safely without friction acting on it and slippery surfaces like ice can pose a great danger. The lessened mobility of the increasingly denser material planes is consistent with the mobility of our perception in our early development. After completing our wanderings we should return to our plane of origin as fully conscious spirits, richly laden with the experiences gathered during our sojourn in the material worlds, which we then leave behind forever.

The power of attraction from the primordial realm at the summit of Creation is able to unite all of Creation in one whole. This is the basis of cosmic gravitation, and the attractive force of gravity is relatively weak on earth in comparison to other forces of nature due to the inconceivable distance of its true origin at the summit of Creation. The uniform cosmic laws, the Laws of Creation, are adjusted to the consistency of each plane, adapted to its forms and conditions, which are rooted in the mobility. These uniform laws penetrate the cosmic whole, in the visible world and also in worlds that are invisible to us on earth, like a gigantic network of animating nerve-strands holding the entire Creation together in coherence and linking with the Creator.

Before God's Will, which is expressed in the cosmic laws, the human being is like a child entering a room filled with books written in many languages. In reality, it is the same book in different languages. A child will dimly perceive some consistency in the books but will not understand the languages. He must break through every intellectual limitation in order to grasp the uniform symmetry in all. We go through different worlds in order to mature through the experience of their individual expression of the cosmic laws, which may then be ingrained in our spiritual awareness.

Here is a further analogy to our spiritual experience in the world. Consider a piano recital being broadcast live on the radio. It is the music that is heard by us – not the radio. The piano is not playing itself either, but the pianist activates the keys. Beyond all this is, as we know, there is the composer who literally 'calls the tune'. The music belongs to *him*, not to the pianist (unless he is the composer), even less to the listener. He only owns the radio, nothing more. It enables him to listen to the music in which the will of the composer is expressed. The music scored or mental notation used by the artist is translated into physical movements, tapped out as vibrations of the instrument that resound in the ear. Excited pressure waves in the air vibrate on a microphone, producing an electrical signal which travels from there to a transmitter, causing an oscillating current to bolt back and forth in a transmission tower or antenna. This current generates an electromagnetic field that spreads out in radio waves to bombard the reception antenna like electric showers. Then the transformation is reversed, step by step, until the brain of the listener registers the music. The signal has changed many times in the intervening processes, whether mechanical, acoustical, electrical or nervous, but none of the individual signals resemble their counterparts in other media, like movies and books. At the same time, the kind of energy associated with the signal also changes with each transformation of the musical information; but despite this the pattern, the relationship of the magnitudes and positions in time of the individual signals – in short, the form of the music – survives intact throughout.

A toddler may wonder why that music is coming from a box, but an older child might know that there is nobody inside the box and that the instrument is not playing by itself. Our rudimentary perceptions grow more acute with experience, until we are able to discern the principles of transmission. Such is the allegory of our wandering spirit in the evolving phases of self-apprehension, of our inner awakening. We wander as spirit germs to absorb the variations of the cosmic laws in the various worlds or planes and thereby attain an awareness of the wholeness of God's Divine Will, and in this maturity we work our way consciously back to our spiritual origin.

Experience gained through the exertion of our will informs our increasing spiritual awareness. The spirit in us is responsible for our personal development. Consciousness expands in line with our recognition of, and our voluntary adjustment to, the natural laws in all Creation. Starting like a child that is developing step by step to know and adapt to its world, true perception grows with true knowledge. Not the shadowy learned knowledge of this world nor the conceited attitude of the over-valued intellect steeped in its own cleverness, but an inner knowing that emerges from a morally improving vision of transcendent truth anchored firmly in Creation. Finally we can draw conclusions as to the nature of the composer by consciously experiencing the music which has emanated from him. This will not be complete, of course, but limited to the music he makes available to us, as his work. Other sides of his personality we cannot know because they do not touch us. A similar situation between the composer and his music, being a product of his creative genius, exists between God and His Creation.

Nothing takes place in Creation outside the perfect laws, which can never be changed, circumvented or overthrown. This is why, when extraordinary events occur, which can still humble our pride and shrivel our schemes in an instant, we associate them with the Creator. We say 'God knows', 'an act of God', and 'God moves in mysterious ways' when we are in states of metaphysical perplexity. Erroneously, we think that the Divine needs to reveal itself through just such magical events that seem to overturn the laws of nature, events that impress us by their incomprehensibility. But miracles are natural happenings, too, to be found in every grain of sand or grain of wheat. A miracle is not a supernatural intrusive action by God in human affairs, but something that operates within the cosmic laws. No exceptions can occur in any circumstances that are contrary to the naturalness of all the other events.

'Miracles' depend on the nature of the power in operation. Spiritual power can speed movement and cause time contraction in species that are lower than the spiritual. This happening is, as yet, something beyond our ken, a 'visitation from above'. The record of history is strewn with examples of 'miracles'. Miracles wrought by Jesus arose naturally from the pressure of His innate Divinity, a power fundamentally different and infinitely greater than spiritual power, but certainly fulfilled in the working of the cosmic laws.

These 'supernatural' visitations are by no means confined to those we call 'believers', although as a result the recipient will naturally become a believer of sorts; they are available to all human beings and are indeed their rightful inheritance. Thus 'miracles' are not simply the prerogative of any particular religious sect, nor did they only take place in the remote past or on the shores of Galilee; they are part of the natural process where luminous rays of help from above penetrate the human soul and are available to all who open themselves by inner exertion to spiritual currents. Of course, some of the 'miracles' reported in the Bible are mere fictions invented by the human brain to 'sensationalize' events or to 'cover up' others found to be 'inconvenient'. Thus it was impossible for the sun to 'stand still' to enable the Israelites to win a battle under the leadership of Joshua. Just as it was impossible for Jesus to be born without a biological father, walk on water, turn water into wine or ascend to Heaven in the same physical body that was brutally destroyed on the cross. The Bible would only gain in stature if these untruths were excised.

FIVE

Time

Our evolution dates back more than two million years. It has endowed our mind with a number of highly dependable senses, and along the way we have become increasingly reliant on our intellect. Indeed, our notion of time together with the temporal order of events is a product of reasoning from given data. But what is time? And does it really pass? In moments of revelation we say that time stands still. Yet we constantly, feverishly one might say, rush into time, muttering that we have *no* time, have not *enough* time, that time waits for no-one, and so on. The concept of time is omnipresent in our lives. Hence such everyday expressions prevail, such as 'time passes', 'spare time', 'free time', 'lost time', and proverbs such as 'time waits for no man' and '*tempus fugit*'. We have become clock-bound creatures, obsessed by notions of having to struggle to fit everything we can into our brief span of life. We look back at history and say that 'times have changed', but time has not changed, because it is eternal. Only the forms change.

Time indeed possesses paradoxical characteristics. It is simultaneously long and short, long because it has been going on for billions of years, short because we feel we do not have enough of it. But we nearly always find time to do something we love and enjoy. We also experience time as slow or fast, a slow drag of dull moments for one who is waiting, but it flies fast by when one is absorbed in activity, enjoying life or, as we say, transported. It can be large, up to eternity and infinity, and it can be as small as the fraction of a second. It is capable of immortalizing men, but it also causes them to be forgotten through death. Time can efface recollections, but it can equally bring them back to the surface. On the one hand it is rare, but we also waste it, together with the resource of opportunity it represents in the passage of our life; on the other hand, we always find ourselves with some time, even while we complain of not having enough. Whether we use it or not, it disappears, seemingly never to be recovered. For some it is a predator, for some it is a companion in the sunlight of an endeavour. For many it has monetary value, hence the axiom 'time is money' and the sentiment of 'buying' time, yet in fact it can be neither bought nor sold.

Our experience arises from the interplay of our conception of space. Thus time passes or flies. This rings true with our life experience – while awaiting an important event that must take place at a certain date and time like an examination or an appointment, for example. This event is far from being a reality, being still remote from the present moment. But little by little the appointed time approaches, by our shrinking or expanding sense of the minute, of the hour, of the day, and so on. The distance between it and ourselves progressively diminishes. It comes closer and closer, becomes imminent and, suddenly, it is there – or, rather, it is here and now. It has entered the present moment. It has arrived.

Within a very short time, however, the event is past; it is behind us. It distances itself progressively from current reality, even sometimes from our field of consciousness, from our streams of thought and emotion. The once important event can now only be recaptured by memory, and soon not at all, the present operating as a reflecting mirror that limits our range of retrospection or how far we are able to look back. Certain periods in the past then appear closed to the mind. Thus we feel that time is like a river, ever flowing, never still. The flood of time advances in one direction only: coming from the future it brings events to our present, only to disappear into the past. As the hymn says: 'Time like an ever-rolling stream bears all its sons away'. The process has no beginning and no end and is repeated indefinitely. As we say, time passes.

Echoes of reality

However time unfolds for us, it does so in a definite, individual context. While clock time indubitably does pass, *real* time cannot pass. If time actually did pass, like an invisible river, we would all experience it in an identical manner. For convenience we adopt a system of measurement, using clocks and calendars, much as we measure a quantity of grain or a distance. But our daily life shows us that time is perceived in a variety of ways. It passes faster or more slowly depending on the person and circumstances. This is true of our everyday experience when, for example, we are absorbed in some activity and are called to a meal, only to be astonished to find that a whole morning has elapsed.

By contrast, time passes slowly during periods of low psychological stimulation, for example, when someone has no interest in what he or she is doing. Time drags when we are waiting in a long line or sitting in waiting; it crawls for the clock-watching factory worker, who may be performing routine tasks or be weighed down by a cumbrous operation, or for the disaffected schoolboy, but it flashes by for the artist, the inventor and the executive.

Yet again, if time really *passes*, then before being here in the present it must have been elsewhere. In the same way, after having left the present, it should equally go elsewhere. But where is that elsewhere? Where are the reserves of time that pass, and in what form are they stocked? And where is time accumulated after its passage? What becomes of it? In fact, time does not come from anywhere and it goes nowhere. Difficult as the perspective is for our clock-adjusted brain to accept, time has no beginning and no end.

This brings us to time travel and the proposition of the fourth dimension. Is it possible to project ourselves from our earth-bound space-time – in which space is extended in time, and time is extended in space – and journey back and forth, from past to future at will? This cherished dream was first purveyed by H.G. Wells in *The Time Machine*, whose hero, an eccentric scientist, builds a machine to transport him into other time worlds, where past and future become present. In reality, we have no power to escape the present in time because the cosmic mechanism does not allow it. One reaps what one sows and this can only be rectified by a new sowing, not by 'going back' in time and 'undoing' certain actions. The concept of going 'back to the future' is (and remains) a utopian dream.

But if time does not pass, and does not simply happen to us, if it does not transport our fate or arrange things by itself and can neither heal nor promise, what does it do? The fact is that time actually stands still. It does not change; it is the same, today, yesterday and in a million years hence. Only the forms change in time. We dip into time in order to draw from the lap of its records the knowledge we need in order to enrich our life. Times change, the hands of the clock move, the sun shifts its position and, over eons, the stars too change. On earth, plants grow, children grow up, mature, age and die. The body changes, yet the self is felt the same always, even as our self-awareness grows and changes and whether one looks young or old. The changes in the activity of cells, molecules and atoms are elements that allow us to affirm that time passes, but what passes is not time but the forms.

It is not that the changes of form are caused by passing time. If it were simply this there would be a place elsewhere where the forms of the past and the future could be found. Not only has this place never been discovered, but no suggestion about its possible location has ever been made.

Apart from time, nothing is motionless in Creation. At the level of the fundamental particle and with the cosmic bodies, and right up to the summit of Creation there is motion. Celestial bodies rotate around their axes, and also revolve around the central axis of their galaxy, which in turn rotates around its own axis, and so on.

With no ebb

In an 'abstract' domain, like thought, movement takes place. Our mind cannot, in fact, stop thinking or wishing for something. Always and without interruption, we desire, reflect, think and aspire towards something. These cogitations lead us to act in order to fashion new objects, new objectives, new thoughts and ideas. The mind is never still, always scheming, always developing, in some way or another creative. The unexerted mind is a dreary landscape.

With the aid of free will we, within natural constraints, may direct our path as we wish. We may throw ourselves enthusiastically into life, into the passions and sensations of experiences, directing our activities towards certain goals and ideals, or we may drift, taking things as they come, be resigned to what we think of as our fate.

The experiences of time which result from the different uses of free will vary dramatically. For the active person, time is too short, the minute or hour passing quickly; for the inactive, time drags. The way we experience time is influenced by our state of alertness and current moods. It is fast if we are vigilant and active, and seems to crawl very slowly if we are indolent. In reality, however, it is not time that moves, but we who move in time.

Knowing the hour, being able to tell the time of day, is not the same as knowing time. The hands of the clock on the dial give us a frame of reference of external cues for events and nobody would deny their usefulness. The smooth running of our lives, from pulse rate to flight departure time, may be organized and adjusted with reference to the clock. We might call this frame of reference 'conventional' time, which was invented by man, as opposed to 'real' time, which is a creation of God. In extremity we become conscious of real time, we need to 'stop all the clocks' as W.H. Auden put it in his poem about the death of an intimate friend.

Therefore two different concepts of time exist: *real* time, which stands still but is relatively mobile to us according to our inner movement, and *chronometric* time, which is in fact not really time, but rather a system of measurement that advances at a fixed speed, gauged by the dial of a clock. The latter is objective in duration, and is neutral and common to all, while real time is something else. Real time is not measurable, but subjective in impression, variable and individually experienced. It is unpredictable and connected with out-of-body experience. One's experience of real time is by its very nature life-changing and spirit-enhancing, rather than simply life-organizing.

These two contrasting time experiences match the two parts of our consciousness: the intellect and the spirit. While chronometric time experience is born of the intellect in our daily sleep–wake cycle, real time belongs to the intuitive recognition of the spirit. The spirit, being immaterial, needs a strong resonance of the body, together with brain activity for anchorage in the material conditions. The intuitive mind and the rational mind give rise to our contrasting experiences.

The brain processes a great deal of sensory information regarding events in the body and the surroundings. It organizes and is meant to pass on perceptions of significance to the spirit for our greater enlightenment and self-growth. The brain has to quantify data, to weigh, analyze, measure, judge and calculate such things as size, temperature, density, speed and distance. Naturally it also needs to place these perceptions in time, and the time it needs for this processing is chronometric time.

While the brain classifies and categorizes perceptions, the spirit relates to the qualitative aspect of experience to empower our mental observation, cognition and reason. It is concerned with the finer qualities of experience. The dogmatic view – so prevalent today – that the spirit is too vague or cloudy to bother about, is a grave error from which our sceptical age suffers. Spirit is an entity concerned with growth towards the ideal. And being encased in what we have always called the 'soul of man', it is essential that it be exerted. The spirit draws from the great motive power in Creation; its currents sustain body and soul.

What we apprehend in inner experience can evoke an inner resonance that chimes with finer reality, and in this inner movement our spirit is able to expand in consciousness. This experience of spirituality in a secular world paves the way to growth in spiritual maturity and greater intuitive perception, thus self-awareness beyond the intellectual limitation of earthly time and space. And this is also how time perception differs for each individual.

The goal of the spirit is to develop its inherent abilities and to make them resonate with life. We as personally conscious spirits are then able to make the way back to our origin, the sphere called 'paradise'. However, in shackling the free spirit through subordination to the intellect, we misdirect its strong currents that should further (rather than hinder) progress towards this goal. For example, to leave this world poor in finer experience sends an immature and incompetent spirit to the beyond. It would need to return and retake the missed steps. This of course is a retelling of ancient story, not merely a Christian one but a universal one. If the atheist chooses to write it off as 'mythology' or 'mere allegory', he does so at his peril.

It is no easy task for us time-bound creatures – ever-conscious of the ticking of the clock and of our own mortality – to appreciate that the time of the spirit is wholly other than that of the brain. Thus a moment intensively experienced is much more profitable than a long period passed mechanically, passively or pondering and, by contrast, hundreds of decades of human history may be, in spiritual terms, very short.

An example from common experience can demonstrate the difference between chronometric time and real time. A man leaves for a week's vacation full of intense and lively events, one following upon the other – he may have fallen in love with a place or person. At the end of a week he feels as if he has been away for a fortnight, a month, or even longer. This recognition is of course subjective, for a week can only last seven days, but his spirit has traversed a much longer and more bracing road than is indicated by the calendar. An eventful period appears longer. As novelists and film directors know, time is counted by the spirit; not by the hour but by the nature of experience. A thousand earth-years can be experienced by the spirit in one day.

Scientists model an entity of space and time, the time–space continuum, in terms of a cosmic framework in which all that exists is located and all events occur. Speculative theory aside, some are even blithely wedded to the notion of one day 'conquering' time and space with the power of human reason. The complacency is time and again shattered in the tears and wreckage of natural and man-made disasters. We are simply, according to one well-known hymn, 'dwellers all in time and space'. What we call 'time' is a function of the periodic rotation of the world that shelters us. If the earth rotated faster around its axis, our days would be shorter. We often say 'how time flies!' yet it is not time itself that would pass more quickly if the earth rotated faster but the speed of revolution and evolution; conversely, if our planet rotated more slowly, our days would certainly be longer. However, since the dilation or contraction of all our means of measuring time are the result of our planet's speed, we would not be conscious of any change.

Time seems to us today to pass faster than in the past, when man was guided by the sun rather than a vast array of timepieces, each dividing the day into regular slices. We live longer than our forebears, are often more affluent, but, paradoxically, seem to have less time. Hence the saying 'asset rich but time poor'. We spend a lot of time accumulating assets which then lie around unused and gathering dust because we do not have the time to make proper use of them, since once purchased, we are already eyeing our next acquisition. Whether it is gym equipment, a musical instrument, a hi-fi or home theatre system, a spa, a boat, a caravan or whatever, we are always on the lookout to 'upgrade' or 'expand' without ever getting full enjoyment out of what we already have. Something about our notion of time does not square with our intuitions. Similarly, as we age, time seems to fly by at an accelerating pace. 'Can this really be so?' we ask ourselves. We experience time in relation to memory, remembering events from our current perspective. In childhood, we can barely think beyond what is now, almost existing for the moment. With memory, not only do we not exist only for the moment but we experience time in a linear progression. Rational notions of time have come to seem nonsensical. As indeed, in cosmic terms, they are in the limits imposed on our consciousness. The measuring intellect is limited to material data, from which it constructs the concept of time. Science *measures* the world, but spirit *orders* the world. The quickening of events we observe today is testimony to the general acceleration of evolution in the temporal world. The cause has so far eluded our understanding.

The luminous cloud

The whole of Creation, with planes all the way down to the cosmic bodies of the physical universe, can be compared to a gigantic electric motor. When current is applied to any electric motor it starts to turn. It stops when the current is turned off. We see something of the movement of myriad celestial worlds and dimly perceive an order in it. Science labels this movement 'energy'; biologists use the term 'life'. It is anybody's guess what this life force or energy really is, or how or why it came into being. It is a given, a primary force, as baffling to the scientist as it is to men of faith, who call it divine power or grace.

If divine power were to stop flowing through the cosmos, all would slow down, and in losing speed everything would plunge into chaos and catastrophe. It would be like a gyroscope, which maintains perfect equilibrium for as long as it rotates at an even speed, but which, on losing power wobbles, loses rotary motion and finally stops altogether. Likewise, an artificial satellite continues its periodic rotation round the earth for as long as its orbital speed is high enough to counteract, by centrifugal force, the attraction exerted by the earth. As the orbital speed drops the terrestrial attraction pulls the satellite back to earth. Divine power streams through in radiation to propel all that exists in motion. Without movement, there can be no cosmos, and therefore, self-evidently, there can be no evolution.

Strive as we will to know why the earth should happen to be the only planet in 'our' universe where conditions support life and myriad adaptations of creatures, we must accept that we have no clue if this is really the case. We are clustered in a narrow band around a cosmically undistinguished planet. Our earth is one of a narrow band of planets orbiting an average-sized star. The human species is time-bound, present for only an infinitesimal fraction of stellar time, and it is but a step in a process that we can at best intuit. It would be a bizarre phenomenon to have the universe all to ourselves, when only a few thousand feet overhead or a few hundred feet underfoot, life for us is impossible. Venturing beyond, we must take fragments of our home climate with us to survive amid the vacuum and radiation above or the fire below. Which of course is not to say that we should not try, should not attempt with all our might and main to fathom Creation; indeed it is a condition of our existence to do just that.

The ability to examine and illuminate is human. We are capable of and need to be perpetually probing, asking questions, inquiring into the world around us and into ourselves, weighing, measuring, constructing and deconstructing buildings, kingdoms, theories, hypotheses. This is in part the work of the restless intellect. However, as we have maintained throughout, there are dangers implicit in overvaluing an intellectually-dominated science, which is but a tool kit constructed because of an authentic longing for transcendence.

We are forever thinking, one thought immediately following or displacing another with hardly a moment to spare. Trains of thoughts and perceptions sweep endlessly across the landscapes of our minds, with a tiny part discharging into words or language. The general belief is that thoughts are generated by the brain and are no more than the exclusive product of the intellect. In reality, it is only because the non-material spirit of the human being is incarnated in the physical body and uses the brain that thoughts are produced. Without the spirit, the brain cannot think. Due to its higher origin, the spirit possesses the power and the ability to press out forms from the substance of material planes.

Thoughts are pressed by the spirit's intuitive volition. The human spirit wishes, aspires and asks questions and in doing so elaborates thoughts. Thoughts are not abstractions but tangible realities. They are, however, not formed of the same physical substance as is visible to our eyes in our immediate environment, but consist of a finer material substance beyond the earthly, which is real and possesses a definite consistency and density. When we think, we emit a vibration that already gives rise to a radiation. This is, for example, reflected in the imaging that can be detected in parts of the brain that are active when we are thinking. Thoughts take on a tangible reality in a finer material domain and therewith a consistency and density. The resulting form corresponds exactly to the nature of the thought. The energy forms, which are as strong and durable as the emotions attaching to the thoughts, can ennoble or devastate. A thought nourishes and strengthens and is nourished and reinforced by its homogeneous kind, issuing from total strangers and out of our awareness. Our thoughts are linked with similar thoughts. We thus influence and are in turn influenced by another person's thought or action. We poison our finer environment by negative thoughts.

We are complicit in the fate that befalls us. We are responsible before the cosmic laws for our thoughts. The insidious pollution (not merely industrial) that surrounds us is not just hard to estimate but also lethal.

Ethereal material domain is an alien and different consistency to the gross material domain of the earth. It is a totally different material species and constitutes the beyond, where human souls may dwell before and after being on earth. It has reciprocal interactions with our earthly world through the forms arising from our deeds. Ethereal density manifesting from the entangled ways we live generally and weighing down the earth like briars is throwing it off a healthy orbit and upsetting its equilibrium. This does not mean that the earth's distance from the sun has been altered or that it has increased in mass but it has more to do with the ethereal or finer material surroundings that envelop every celestial body. Invisible to the physical eye, these surroundings, including the consistency that harbours our thought forms, are adversely affected by the negative and base thoughts emitted by the bulk of humanity. A person too can be weighed down and become unbalanced if he is plagued by sorrows or succumbs to negative thoughts. If he is unable to master the situation, sooner or later his physical body will also be affected and become ill, without there necessarily being a physical cause for the illness. So it is with the earth. If these transitional planes did not exist, a human spirit could not be active on earth at all because the gulf between the spirit and the earthly domain would be too great for an incarnation to take place.

We are connected with other people who have similar thoughts and similar emotions, mutually reinforcing one another as though through gossamer threads of finer material substance. The thought works on within the finer intensification, and presses for realization in the visible world. Thoughts can, in their self-acting and far-reaching process, bring about upbuilding or decline. But despite the extraordinary energy of aggregated thought forms, we are not simply at the mercy of their influence without protection. It is enough not to nurse similar but opposite thoughts. Connection can, for example, be made with centres of positive, ennobling and constructive thought forms. The reality of thought forms alerts us to be careful about the thoughts we generate and nourish. The influence our thoughts have on ourselves and on our immediate surrounding is enormous. Life on earth is getting more frenetic. Evolution is speeding up and we today reap the consequences of our deeds more rapidly than ever before.

If every advance in science were matched by growth in spiritual awareness, equilibrium would be maintained. Of course, people have always suffered from extraordinary events, both natural and human disasters; there have always been cosmic collisions, wars, oppression, suffering and starvation. But the revolutions of modern technology, on which we have grown ever more dependent (if the computer 'crashes' everything grinds to a halt) lead, almost inevitably it seems, to a slackening of the spirit. The individual will becomes crushed under the wheels of technology and the growing authority of proud science is allowed to obscure meaning and higher recognition of experience other than those accessible to the intellect.

Springs that never freeze

As one evil counteracts and destroys another, so fire drives out fire. Hence the parable of the Last Judgement in Christian symbolism, where all that is dead is resurrected and all that is without movement is animated.

The world-view of the purifying power of fire or the Light that is heat is not confined to Christian doctrine or religious confession; the idea is universally felt in the human soul. The concept that ultimately nothing is lost, that nothing can be hidden, persists in all religions. There is a God and there are higher beings who know better and, literally, who know infinitely more than humans are capable of knowing. From the simplest plant to the most complex animal there is a movement towards the Light that animates and informs all.

That which does not strive towards the Light and which cannot adapt to its intensifying pressure will be consumed. So it is with human souls. What has hitherto appeared to be dead, what has slumbered in the soul, often unknown to the person him- or herself, will be roused and be strengthened. What is adjusted to the Light, to the cosmic laws (the Laws of Creation), which bear the Divine Will, will grow in the purity of its volition. The active forces of evil, human hatred and exacerbated passion will all be spent in the Light's power. These souls will literally run themselves into the ground.

So it is that our further development is our spiritual and moral capacity. Whether we realize it or not, whether we accept or deny it, we are at a new dawn of evolution, in the direction of spiritual consciousness. As it is, we see our future evolution, albeit through a glass darkly, as still free and open, but it can only be hampered by excessive reliance on the powers of intellect and technology. It is our task to press the powers of the intellect and the technology issuing from it into the service of the spirit by pursuing spiritual goals. Technology should never be an end in itself and things should never be produced just because it is possible to produce them, merely to satisfy our vanity. Thus we have succeeded, among other things, in producing chemical elements that do not exist in nature at all, such as plutonium, which is not only a radiation hazard but is highly toxic as well. We now meddle in the genetic structure of living things that were built up during millions of years of evolution, according to the laws ordained by God, as if we can 'improve' these structures, thereby denying the perfection of God and His laws. Future generations will shudder at the boundless conceit displayed in these foolish attempts.

We human beings exist in a certain accord of spirit and body. Our core is illuminated by a spark of living spirit, which has nothing to do with intellect or reason, and is cloaked outermost by the physical body. Spirit cannot be studied by brain or thought activity; it works above the plane of thought. Instead of sharpening the intellect, we should be sharpening our intuitive faculty, being the voice of the spirit that works through the solar plexus and the cerebellum. The cerebellum should be gradually restored to health so that it can again fulfil its function as a necessary counterweight to the frontal brain.

The human spirit possesses the attribute of free will. It has the freedom to choose any path it will, to strive in a direction of its own choosing. Consider how we are influenced by the thoughts of others, yet we have the choice to attend only to that which we wish to be part of. For instance, when we have negative or destructive thoughts, then we are connected with other similar kinds, that is, bad thoughts in others echo in us. We have the capacity to indulge the kind of thoughts that we ought to have, and determine our associations.

We human beings can be enslaved by our thought habits. In extreme instances, accustomed lines of thought or an inclination that predetermines action restrictively (such as being overawed by passion) make the existence of free will seem questionable. The spirit is hindered from controlling and directing the hidden power of thought by restrictive, often compulsive ways of thinking. The urge rises in an awakening spirit to shake off binding ways of thinking.

But thought control is not achieved by thought. Thoughts are not overcome by thoughts. Trying, for example, to use meditation to command thought with thought, or to limit, end or steer thoughts towards certain ends will not control thought, because focusing attention on any thought merely strengthens it, supplies it with spiritual energy, like everything drawn into the light of consciousness. Controlling thought effectively requires exertion of the spirit, because thinking depends on its power.

A dome of many-coloured glass

When one comes to speak of 'spirit' or 'soul' it is wise to remember that we are using *language*, a strictly limited offshoot of the brain, to talk about intuitive matters that are not amenable to mathematical proof or linguistic analysis. In other words, one is attempting to name the unnameable. This caveat should perpetually be borne in mind when we attempt to account for the unaccountable and to describe the indescribable. After all, language is but a tool, the vehicle of rational thought – useful though it is – and almost, for humans, indispensable. Thus we are obliged to speak of 'soul', 'spirit' and 'the Creator' as if they were as tangible as 'door' or 'dish'. But as with abstract nouns, they exist as feelings or intuitions. For us humans, therefore, they need to be named, and those who know what we are talking about will know what we are talking about.

In this discourse, we are in one sense attempting to rewrite those spiritual laws that have been with the earth from its 'beginning'. The reader may experience some difficulty in 'swallowing' all of the arguments propounded. This difficulty, I submit, may be directly attributed to many centuries of overdependence on the power of the brain, on so-called 'logic', and something called 'common sense', which all too often masks spiritual indolence. The common sense to be applied here is the sense we all have in common, the ultra-sense that there is infinitely more to life than matter.

Many phenomena defy physical explanation. The laws that govern finer material substance in detail, even if they cannot be described by modern physics, can nevertheless be logically experienced by the human spirit; in other words, they can be fathomed intuitively. Ethereal phenomena are not something peculiar and separate from the world of matter, but are part of one great reality. And because the Creation is, by definition, *one*, we find connection between the material and the spiritual. Hence, the human need to speak in metaphor. The vulgar (and, for our purposes, the simplistic) division between the 'literal' and the 'metaphorical' dissolves when we come to speak of spiritual essences. There are no formulas and there is no language that will prove entirely satisfactory to our logical brain-dominated selves.

The human soul, at its core the spirit, is unique among all the creatures. The human spirit is able to develop a personal volition. Humans do not swing unconditionally in the cosmic laws. The animal soul is of a completely different species that enables it to draw life and strength. The animal adjusts unconditionally to the natural laws, unless impeded by human interference. It is without free will and simply adapts in its activities to the laws, even when they are there to keep other creatures on the alert and in motion.

The animal soul forms itself from a group soul of animistic species, to return there again after gross material departure, attracted by the aggregation. These group souls are not so much a formless mass, from which parts taking on form sever themselves as souls, returning to lose their form and merge again with the mass. Rather, they are indistinct formations, intermingling, interlacing and perpetually changing in the course of development of the different creature types.

Only those creatures that are able to move about, such as animals, possess these souls, through which they have the possibility to protect themselves. Plants and rocks do not possess a soul of their own, but form dwellings for beings that have taken on form, such as elves and nixies and so on, which can move out of them as they like, without suffering if their dwellings are damaged or destroyed. These nature beings tend the dwellings given over to them, which cannot move of their own accord.

Our natural selves

The relation between body and spirit, between matter and living entity, is not always easy to grasp. Sometimes the 'fit' between cloak and spirit seems incongruous. Thus crude physical appearances can wrap around delicate natures, and vice versa, such as in the fairy tales of *Beauty and the Beast* and *The Frog Prince*. The body is a vessel that enables the spirit to gain experience. In dreams and fairy tales, the disparity between appearance and entity is often apparent. A dreamer may thrash about, weep or scream, convinced that he is undergoing physical or mental torture. It is as though his body has been literally possessed by an alien spirit, with which, on waking, he feels no essential connection. Only on waking up does the realization dawn on the sleeper that the experience was of a different nature, despite the presence of tears and other bodily evidence indicating that an infiltration by the intellect occurred.

Without the body, the spirit would have no bridge to the earthly. The goal of incarnation – certain development of the spirit through experience – could not otherwise be realized. Remember that we are dealing with matters that are not subject to proof beyond that of undimmed intuition, or our effortless perceptive capacity for immediate insight without observation or reason, because theories and our verification or rejection of them belong to the reasoned analysis of the intellect and remain within the boundaries of gross matter, which is the sphere of action of the unaided intellect. If we are to explore a large mansion or manor house without having access to the floor plans, we must laboriously explore it room by room and then try to figure out the purpose of each room, as well as its relationship to the other rooms. We can then draw conclusions, true or false, as to the nature and purpose of the whole structure. This is how the intellect works. If on the other hand, we had access to the floor plans as well as to the intentions of the architect, we could grasp the 'whole' quite easily and quickly without having to collect and evaluate large amounts of data. In relation to Creation, this second, intuitive, way holds far more promise of success than unconditional reliance on the first. This was realized by sages and prophets of all ages as the only way forward. The connection here is that the argument to rely on intuition rather than intellect alone to gain insight is at least as sound as that propounded by Genesis or any theory of evolution, perhaps more so in that it excludes neither absolutely and is inclined to reconcile both.

The eye made quiet

The spirit leaves its origin unconscious, spurred by an urge to develop to a personal ego. On each plane downwards, the spirit takes on a cloak of the substance of that plane as a matter of course. In the vicinity of the earthly, it takes on the astral body. We appear on earth in the physical body. The astral body is the garment most intimately related to the physical body, and it is the prototype for our physical body. This in part explains the phenomenon of 'phantom pains' experienced after the loss of a limb, for the astral counterpart limb remains intact.

The spirit's volition glows through its telescoped coverings and, causing movement, can shape noble forms in the finest consistency of the plane. Each human shapes his environment by the nature of his volition. No two people can tread the same path. The motives governing their behaviour are too varied for this. One may take a steep path and arrive quickly, while another may prefer a more comfortable route that takes longer. The difference lies in how each, in his free will, travels the road, how his developed personal abilities can help him along. Our psychic development has accustomed us to think metaphorically of our progress through life as an upward path, one that implies struggle and difficulties en route. A further metaphor for the object towards which we are striving is that of light. Using these two metaphors we can say that every person will, by his own volition, take his own path towards the Light. But our human longing for light is no mere metaphor, just as, non-metaphorically, each plant will incline towards the sun. Such is the nature of living things on earth.

Both people and plants are impelled upwards by their life force. For humans, it is their spirit core, by which they aspire to free their will from the shackles of the intellect that – through a decision made long ago, even in the course of a number of earth-lives – they allowed to dominate. We have the free will to use our intellect when we are active in spirit, or we can let it run amok if we are not. Intellect is baffled and frustrated when it attempts to get to grips with spirit, so much so that it often prefers to ignore or even deny the spirit's existence. This is the most grievous error that we can make. It leads to a myopic view of life and its purpose, and is responsible for the crises of conscience.

We all carry images imprinted inside us – certain conceptions of our environment, the people around us and our opinions of ourselves. We also carry opinions with regard to the Creator and His working. They shape into attitude and are, to a certain extent, historical or may even be confessional, but they are hardly scrutinized, except in the most lazy way by simply denying the existence of anything beyond the physical. These backdrops to our inner world affect not only perception but mental well-being. Nobody should in their heart of hearts be really indifferent to whether or not there is a God, but many people do not really seem to care anymore. The way and manner in which each individual answers this central question completely changes our outlook in fact, on our surroundings, as well as the way we perceive and conduct our own lives. Our relation to God decides in a fundamental (although often unexamined) respect, our motivations and actions, so that those who have discarded this anchor of a belief in a supreme being simply out of laziness, appear as flotsam and jetsam, following this fad then another, sauntering aimlessly through life, seeking only earthly recognition and pleasures and a comfortable life. The intellectual scrutiny that may end up as atheism lacks the grasp of undimmed intuitive perception of the perfection of the cosmic mechanism and should not be confused for an inner exertion. It can be due to none other than spiritual indolence. As Wordsworth observed 'While with an eye made quiet by the power of harmony, and the deep power of joy, we see into the life of things'.

Some people, for example, submit to atheism because the dogmatic conception of a miracle-working majestic figure who sacrifices his son makes no sense to them – the ancient and outmoded notions behind the act of sacrificing one's own child or unenlightened religious observances are not thoughts we are likely to contemplate in this day and age – or because of the injustice and suffering that is so prevalent in the world. Someone may dispense with belief in any higher justice because he himself has had a difficult life. Or against the background of his own alienation he no longer asks whether there is a 'higher will'. He may surmise that life is an uncontrolled, unbidden game of chance, and so adapt to the thoughts of others, all trust banished. On the contrary, the believer who subscribes to the traditional pictures of God is certainly on a different path, but no less imperilled. The conception of the punitive, vindictive judge, like that promoted by the Old Testament, for example, is naturally subject to hidden anxieties or guilt feelings, and is indulgent of timidity and spiritual paralysis.

The alternative picture of a saccharine Creator who unconditionally loves us all as children and whose affection we can gain solely through confession is, in truth, equally numbing. Whoever imagines himself saved because he attends church dutifully, confesses, or reels off verses of prayer is in a grave error. Such complacency hardly gives impulse for looking for tasks and meaningful connections. The spirit is pushed aside from its nature, and it will not be surprising if it becomes feeble and indolent.

Our own conceptions of the being and working of the Creator can become quite dangerous for us as soon as they fall prey to devaluing, fear-inducing tendencies. It is the same with our pictures of the world, even as we assume that the world is as we see it or that others may be biased if they do not see it as we do. Whoever assumes that everything is incurably sick and the whole universe is in the end only an accident of great overpowering chaos will not come from this devaluing view to seeking truth or meaning. Whoever daydreams only of the apocalyptic end of the world, expecting great disaster, or at least the ruin of the economy and society, will in his penchant for menace make heavy weather of recognizing the many life-affirming motifs or opportunities that each day offers. And the utopian likewise lays up trouble for himself.

Devaluing, menacing and inadequate are the conceptions we habitually make of God and His Creation, in the end never getting away from a human basis. Unsurprisingly, the pictures we draw for others can also exhibit these dangerous tendencies. Whoever depreciates or despises, perhaps even hates or loathes himself and others, is obviously a cynic and a pessimist. In persistently devaluing himself and the human image it will hardly be possible for him to discover in others or in himself useful abilities worthy of development. In the same way, someone who lives in constant fear and with the tendency to see in others wickedness, menace and horror, will steadily destroy his own inner security by fear.

The stars that shine

Yet ever-present help surrounds us in the working of the perfect cosmic laws woven by God into His Creation from the beginning. By observing these laws we vaguely perceive a glimmer of His stupendous magnitude. No need is so great as to be without hope, no pain without a blessing! The only times we fail are when we want to force our own will, without regard to the eternal order, and when we do things in half measures and rely solely on our earthly intellect. But we choose to stay tuned to the confused noise of the world and do not hear the awakening voice of the spirit. It speaks to us deep within, having first to penetrate many walls, to surmount many obstacles, before it can reach our ego. But it does not give up.

Nevertheless we are in a quandary, not knowing how to access this help that is so close to us. We do not even realize that we can do so by prayer. Nor do we really know what it means to pray, even when urged in times of great joy or deep pain. Praying is not about pleading and begging in anxiety and need, or habitually babbling empty phrases in dull submission. In a serious and heartfelt prayer, a human soul opens itself before his Lord and Creator, kneeling at His feet in the greatest fervour he can command. He is uplifted by virtue of the magnetic power of the ray of love that has flowed to him. New life and new energy flows through the spirit that looks up to God in humility.

Gratitude for the experience of love inherent in the automatic activity of the cosmic laws is a channel to knowing the real power of prayer. Help can reach us by adjusting to the Will of the Most High manifest in the cosmic laws. And even if we bring forward a particular wish in prayer, benefit can only reach us lawfully in return through reciprocal action. To kneel down in wordless, fervent prayer before the greatness of God and beseech Him in a boundless longing to bring to deed the object of our prayer according to His Will, is to get to know the blissful answer that comes from on High when the supplicant is selfless and humble.

He who has never experienced within himself the glorious majesty of God and the grandeur of His Creation, he who has never perceived his own smallness deep within, does not yet know what it means to pray: a profound self-surrender, a single jubilant soul seeking to connect with God. The power of prayer is transforming. Luminous threads are attracted to the person who seeks this connection with the Light. These threads can reach him finally and become a firm anchor he can hold on to in every distress. He must then tend and maintain the path that leads to the Helper, that is, to the recognition of God. The way there for the human spirit is a voluntary adjustment to God's Will. It is like an earthly path: if we travel on it every day, it will become familiar to us. It will begin to appear shorter and ever easier, even in the dark of the most excruciating experiences.

A tremendous power lies in the inner life of the human being! If it is developed and used for the good, it uplifts our surroundings; if it is darkened through pursuing base aims, it presses the surroundings into the depths of darkness. If humanity were united and completely filled with good volition, then the earth would rise upwards again to the place the Creator had chosen for it from the beginning – but from which humanity choose to stray.

So great is the power of prayer that it can open gates to higher realms and draw on the power of the Light that is God. We can wrap ourselves in this Light energy and protect ourselves against base inclinations and temptations. If we live completely in it, we will transform everything around us. How often do we struggle to help somebody off the wrong track? Often, this person closes himself to us. He thinks he is misunderstood and misjudged and does not see his mistake. His intellectual reasoning hinders the recognition of his spirit. If we turn instead, in prayer, to his spirit, and ask for help, it is wonderful to experience how the person, himself unaware, can be transformed. A web of luminous threads envelops him and keeps debasing forms at bay. One inhibition after the other, to which he is still clinging erroneously, can fall away from him under the influence of this energy of the Light. He becomes brighter within himself, because his environment has grown brighter. We often feel an inner radiance emanating from a complete stranger, whose language we may not even know. We understand, as it were, without understanding. The heart understands better than the tongue. We are in receipt of psychic radiance, popularly known as 'good vibes'.

The human soul can acquire qualities by its own choice. Uplifting qualities are empowered and upheld from fine, higher realms in the Law of Attraction of Homogeneous Species. Dark qualities, possible where the spirit has grown indolent and is walled in, cause the coverings around the spirit to become dense, dimming its radiance and dulling intuitive perception. The environment of every human being is permeated by his or her ethereal affinity. Good will counters ill, and either absorbs it or weakens and even eliminates it. Good will, like ill-will, will have its intended effect only where it meets a homogeneous soil in another person. But those who do good from motives of vanity or ambition do not give off positive radiation. The motive is all, the spirit behind the action.

Footfalls down the passage of discovery

As in nature a plant seed rooted in earth thrives amidst friction from diverse forces and elements, so our spirit seed gains from the friction of its interactions with the material world. The spirit awakens to conscious activity, stirred to decision-making by the vibrations of development and decay that pervade matter. Just as the seed needs the elements, our spirit needs the struggle that comes from friction in order to progress.

The human spirit, by virtue of its higher species, exerts a force of attraction on material substance. It rubs constantly against its environment. This challenges it to be alert, to assert itself and be decisive and, in the process, bring to their full potential the abilities lying dormant within it. These abilities develop all the more by being used. The environment constantly spurs the spirit to a wish, for example, to be loyal, helpful, friendly or respectful, which motivates the corresponding faculties into action; and the more the abilities are used, the more they develop. Its ability of free will is a balance for the ability to attract. It enables awakening of sensitive volition. Otherwise, the spirit would attract indiscriminately and produce disharmony everywhere. The spirit attracts that which accords with its own volition and striving. It may 'sample' here and there, rather like a bee in search of nectar, then settle in a direction and determine its surrounding. Free will makes us attract or be attracted by our choices. Everything therefore befalls us with the impetus to our inner movement and development.

The spirit itself possesses only pure and noble qualities awaiting development. It did not create itself, or the potential abilities within it. Of course, not every spirit will automatically germinate towards the goal of development. Its pure qualities can be cultivated and brought to fruition, or they can be buried with the undeveloped spirit by dulled coverings. Our spirit germ encounters temptation once it incarnates and struggles to resist the possibility of yielding to weakness. The principle of temptation, however, was not willed by God as an instrument of development. It was introduced by Lucifer, originally sent out to support the developing spiritual, in his endeavour to 'speed things up', but it had disastrous consequences both for himself and for the developing spirituality on earth. But this does not absolve humans of their responsibility because there was no need whatsoever for anyone to succumb to these temptations and take the wrong path. Lucifer could only entice – nothing more. He could only do harm where he struck a similar chord. Humans have free will and cannot be forced to do anything against their will. If they fall it is because they want to, and if they rise again then it was their volition that preceded it. In the material planes the spirit germ can sully its coverings through indulging in base vibrations, which become superimposed upon the noble characteristics.

All guilt and karma exist only in material planes. It cannot become *part* of the spirit as such, but only cling to it. Every bit of impurity must be dispensed with and cast off before the spirit can be liberated for its return to its plane of origin. A uniform maturity of all humanity is not possible. The ranks of humanity in any period will always show a great variation in the state of experience, since our development is personal and will always depend on how we exercise the free will that is intrinsic to our spiritual nature.

Light

Today, people do not seek God but secular authoritative voices for the answer to their quest for meaning. Secularism is the norm in the shift from religion and metaphysical claims to reason. Moral and speculative philosophies and value systems have mostly been drained of religious vitality. The scientific concept of the universe as a self-actualizing mechanism dispenses with the need for a God to illuminate and order the world. We should be concerned with the natural order and nothing else, for all else is unknowable and speculative clutter. Such is the accepted world-view in an age of secularism. The paradox is that in the value-free ethic of our age of lost faith we are as subject as our ancient forebears to impersonal forces, and as puzzled by them and by ourselves. Science has provided many answers, but each answer opens a Pandora's box of further questions. Today we are seemingly subject not to the Will of God but to bedazzling capricious forces in a universe with no purpose, to chance, chaos, and sundry depressing ideologies. We have 'conquered' time and space but cannot predict even one event with certainty, except perhaps 'death and taxes' as the old saying goes.

However reluctant we may be, out of pride or stubbornness, to bow to a divine ruler of Creation, a sense of the motivating power of some organizing intelligence fills us in every thought, deed, aspiration and longing. There seems no good reason to jettison the time-honoured name of 'God' for the source of this power. We cannot correlate God's providence in the world with His otherness outside time and space. The entire Creation depends for its existence upon the renewable power of God. The universe, like the entire Creation of which it is a part, did not come out of nothing or nowhere but was willed into being, created in the out-stream of living radiation originating from God. Living things have not originated from inanimate matter, nor did life arise out of a complex chemical brew or some such primordial soup. All that exists takes on form from the power of God. In this is reconciled the Old Testament story and the Theory of Evolution. For development too is God-willed.

In speaking of God, it is impossible for us to conceptualize a being that is unsubstantiate and uncreated. As described before, the term 'unsubstantiate' is an expression of our language for something which really cannot be grasped by the human mind. All our attempts to visualize or describe God's otherness must fail. God is Light, outside all concepts of time and space and without relation of precedence, yet His Power impels all movement, all objects and all thought. He is none other than the mighty Power of Life, which gives origin to all existence and animates every creature. The fact that through the ages God has been cast in myriad imaginings, and the fact that we in the western world have habitually personified Him using the masculine pronoun, simply bears witness to human inability to comprehend the divine that lies above our species, and more importantly, outside Creation. There is no shame in this. A greater shame would be to take for granted or ignore the power that shapes all things great and small, all things dull and ugly to our eyes perhaps, but still bright and beautiful.

However, we can see in the example of Christ the way in which the divine touches the human, using an emissary. To show the way to an erring humanity hopelessly entangled in matter, God incarnated a Part of Himself in Creation by a process of unbreakable radiation in a creative act of direct revelation of His Will. This is the basis of the absolute unsubstantiate otherness of Jesus, totally other than a human spirit or a man deified. Other faiths may, with justice, find resonance of this universal message, together with its glimpse of transcendence, carried by sages such as Moses, Krishna, the spiritual teacher who allegedly recited the *Bhagavad-Gita*, Lao-Tse, the Buddha or Mohammed, who brought the divine message through the *Qur'an*. Each of these messengers points the way to human recognition of God's will and power, paving the way for men to absorb the Christ Spirit, for 'nobody comes to the Father but by me'. This does not in the least mean that the Christian churches have 'exclusive rights' to this path.

Great distances below the spiritual sphere – where man has his origin – lies the ring of animistic substance, the dwelling place of the last intrinsically mobile species downwards in Creation. Lowest in Creation are the material spheres, first the ethereal domain and lastly gross matter, the coarse material planes to which our planet earth, together with the physical universe, belong. Both the ethereal and gross material spheres are not intrinsically mobile, but have to be driven by the animistic and spiritual forces penetrating from above. Both material species provide coverings for incarnating entities such as animals and humans, and also 'workplaces' or 'fields of activity' for other entities such as gnomes and elves, right up to the 'gods' of antiquity, whose sphere of activity comprise whole planes, the highest of these beings being responsible for the formation and development of matter as a whole. All Creation is one great reality of radiations and the diverse union of radiations that gives rise to different forms. The universe itself, from subatomic particles to galaxies, evolves out of condensations of cosmic radiations. Man on earth stands enmeshed in a vast loom of radiations of innumerable hues and tones. His volition takes shape in the resonance of radiations.

In the eternal cosmic cycle, the rotation of the material domains, encompassing imponderable planes far beyond our universe, reaches in each part of its cycle of evolution the point of dissolution. Matter decays and recycles, liberating enormous energy. The scale of the event takes in decay of radioactive elements as well as black holes. Fresh union of energy leads to a great new cycle of material forms, with fresh power and virgin soil for development.

Gilded lilies

Each human spirit is attracted to incarnate in an appropriate body on earth, in flesh, bone and blood in a particular culture with its own mores and beliefs. The ability of free will to decide means we can develop personally in a great variety of ways. Each step on the journey of development must be thoroughly experienced before the spirit can advance to the next. Spirit determines the great variation in our collective experience in any period, and shapes the race and cultural life on earth. Culture represents the shared aspirations of a group. Each race has certain abilities and values that other races do not have, or have not developed to a similar extent. Each is part of the single human race, and each, while striving to develop its own unique identity, has something to learn from the others. No one race or culture can, on its own, advance the world or humanity, but the races, just like the cultures, complement each other. No culture and no civilization can thrive if it does not exist in harmony with the world. Cross-fertilization, in people as in plants, is essential for healthy development. The strongest plant is the hybrid, and the most vulnerable animal is the thoroughbred. But a plant, if it is to thrive, also needs stability. A wholesale root-and-branch transplantation to alien soil weakens resistance to unfavourable influences. Similarly with humans, the indigenous soil is the most likely to produce the richest fruit. Absorbed influences from afar should be transformed to provide a sense of indigenous belonging and rootedness.

In other words, a balance must be struck between the *absorption* of the alien and being swamped by it, even amid a social hunger for change. And it is not simply a physical or psychological shift into alien soil that risks a dangerous drawback – it is the intruder who is in alien soil. This is seen in the undermining of indigenous cultures of the New Zealand Maoris, the Native Americans and many other minorities throughout the world. It may be worth noting that as a reciprocal effect it is just the former imperialist countries such as Britain and France (and their derivatives such as the United States, Canada and Australia) who all have much to answer for regarding the treatment of their subject peoples or indigenous populations; they have for some time been the target of massive immigration, both legal and otherwise, mostly from their former colonies or countries in which they have made military incursions, both in the past and the present, with accompanying strife and unrest.

Many so-called primitive peoples are not just strange specimens of frozen human development, interesting fodder for antiquarians, but regressions from former advanced cultures. Works of art that adorn museums and galleries, rustic ruins and archaeological sites testify to enduring spiritual legacies that can still inspire long after the race has vanished.

Harmony does not reside in uniformity, but in complementarity of diverse forms. There are no 'superior' races; each race makes its contribution to the entire humanity. Empires rise and fall, and civilizations perish and are reborn in different forms under new conditions. Thus no race should attempt to imitate another. To ape another culture is to become a caricature and to starve native roots. Thus, for instance, there is a certain uniformity of dress throughout the world, while the native costume is reserved for festive occasions and folklore presentations.

We can learn much from bygone cultures. In fact, it is our sacred duty to do so, to move forward in our spiritual development, even as we look backward, striving to stay alert in our earthly surroundings. Any stoppage, any shackling to what exists, or to the false gods of science or superstition, is standstill, which will surely bring retrogression in its wake. A culture that holds only to its past must vanish.

Scientific investigation these days moves at such a pace that it is difficult to keep up with its latest developments. And the speed increases daily, from heart surgery to cloning, from the discovery of DNA to photographing galaxies. But unless spirit develops to dominate and lead the tool of intellect, disaster looms, for undue materialist striving ignores and undermines our spiritual values. For all our sophistication, the parable about the tree of knowledge in the Garden of Eden simply gains in potency as the centuries pass. What is earthly should not be rashly seized, nor prized unduly, and earthly knowledge must go hand in hand with what is spiritual, which is *real* knowledge and only that leads to the proximity of God.

The vale of soul-making

Diversity drives nature. No tree is exactly like another and no animal a mere replica of its parents. And the higher we go up the scale of evolution, the greater the variety within species and the greater the complexity of the organism. However, physical attributes of race variations do not determine the inner self. In fact the inner life determines the biology. The colour of skin by itself says nothing about personal character. Despite our tendency to foster 'blood relationship', for families to stick together for mutual support, and for parents to take pride or otherwise in seeing their own or their family's traits in their children, the discovery of spiritual affinities is far more crucial to our progress.

We differ only minimally in genetic diversity. Of course there are differences among us, in the same manner that any two individuals will differ. But no DNA code is exclusive to one specific race. The variations from person to person reach beyond racial typing, and the few genes that occur more frequently in one particular race or ethnicity are by no means shared by all the individuals. Only a tiny proportion of our genes varies between races, therefore race may be seen as a cultural cloak, one that shields rather than reveals the true nature of the individual personality. Genetic variation between Africans and Europeans is only a little greater than *among* people of either race. Just as Africa is the most ecologically diverse continent, as well as being the cradle of our evolution, so African people as a whole show greater genetic variation than the rest of humanity. Thus to talk about an 'African man' or a 'black man' as if he were a racially different or even in some way an 'inferior' human type, is nonsensical.

Despite the popular prejudice played on by sundry dictators, and genocidal ideologies that want to dominate the globe, any attempt to rank humanity on racial or biological ladders is futile. The intimate revelation of the self lies far beyond the province of genetics. We are much more than our genetic makeup, just as we are more than the product of our cultural heritage. We develop in our freedom of volition.

Necessary journey

Spirit is inherently free. It is not bound to race, culture or to the body that gives it shelter. But spirit also has the task to develop, without becoming earth-bound. Before the present incarnation it has already been on earth at least once before. The dynamic of reincarnation places race and nationality in a limited context. Reincarnation is a means to the end of development of our spirit core on earth. German Nazis may have reincarnated as Israelis, and Jewish Holocaust victims as Germans. Black people may be born white next time, and vice versa. Spirit reincarnates, and every human being has a responsibility to develop in spirit. We should bring to life in the expressive power of our nature the abilities that slumber, as it were, in our spirit, for our own self-development and the ennoblement of our surroundings.

We are free in spirit to make decisions of our own, but we are unswervingly subject to the consequences of our decisions. Whether the decisions are from independent volition or the result of the influence of others, the consequence is the same – some major or minor shift in our personal circumstances or in the balance of earthly powers, some greater or lesser vibration which will have its inexorable effect in Creation. Thus we see that this earth-life is but a tiny segment of a person's entire existence, taking the word 'person' naturally as encompassing body and spirit while in the earthly plane, the body being but matter to be dissolved into its elements after the departure of the spirit.

During the brief passage on earth, our will is free or is bound to any extent from earlier decisions. Its task is expression of the spirit on its journey, a journey whose ultimate end is that 'undiscovered bourn' – that unknown country of death – that so puzzles the thinking brain. Contemporary man, fettered by intellectual domination, accords little or no relevance to the immaterial. Discussion of God is often viewed as a solecism, and the concept of a spiritual entity is seen as outdated or even childish. 'The limit of my language', says Ludwig Wittgenstein, 'is the limit of my world'. But there are authentic ways of knowing that are beyond intellectual patchwork and beyond words. This holds especially true with regard to what is most sacred in us, and what must be fully grasped. Einstein, in describing the insights to concepts such as his theory of relativity, declares that 'words and language, whether written or spoken, do not seem to play any part in my thought process'. We are more than flesh and blood and the world is charged with spiritual essence, whether or not we open ourselves to its vibrations.

Our intellect is directed singularly towards earthly knowledge and power, hostile and obstructive to everything that lies beyond earthly comprehension. In exaggeration with our inner indolence it cuts us off from comprehending all that lies outside the physical world, thus shackling our spirit. Since God is utterly outside the range of intellectual perception, the intellect is clearly an obstacle to the recognition of God. It is by its nature averse to all that is divine and all that is spiritual, simply because it has no conception of them. The same limits are imposed on all those who are subordinated to their intellect, and seek to build their lives within its context.

Certainly there are elements of Christian belief – to take but one of many attempts to explain the deep truths of life – that are hard to reconcile with the hard reality of the world as we creatures on earth experience them: the justice of God, for example, in a world where one person lives in luxury and another in poverty and hardship, one is healthy and another crippled from birth, one is fertile in mind and body while another is sterile or wasted. And fate strikes some so cruelly, while others seem to have 'all the luck'. The cynical will just shrug and likely say there is no justice, the believer will dwell upon mysterious ways of God and hope for or trust in compensations in an afterlife.

In fact, birth on earth forms only the beginning of a special section of our entire existence. Life here and now is not the be all and end all, and not our very beginning either. On the question of justice, for example, from where did the idea come? The fact that we question the ways things are, giving moral depth to our perception, that we realize that we can in fact use our will to make life fairer for the inhabitants, suggests that someone or something is active in our consciousness, that there is a will to change things for the better. From where did that will come? It was not in our genes and nobody told us about it. Could it not come from the creative will that allows us to develop, to choose one course of action rather than another, to develop not only our mind but our spirit, to cooperate with the intelligent order underlying all matter and all thought; ultimately to cooperate with the mechanism of Creation in the great God-willed development as a fully conscious and mature personality?

Since it is all but impossible to speak of the immaterial without using material images, let us take the common image of the thread, particularly that of the threads of fate. Every thread of fate is ethereally formed through an act of will. A thought-form may travel any distance from us, but it nevertheless remains connected to us as the producer through the fine material thread. It is the existence of such threads that enables us to recall thoughts. We take a certain decision and strengthen this or that thread, adding and knotting new threads to the whole every day, every hour and moment, creating them by our will. The karma or fate that we bear with us is a result of existing threads we have formed from our deeds, reinforced by other homogeneous threads. Thus there is reciprocal action, the past working with the present, the spiritual experience with the earthly will.

Unless we recognize the spirit's reincarnation we cannot comprehend the flow of our existence nor the self-evident deep patterns and truths of life; we will only see what lies on the surface of our existence, rather like a mariner who judges the size of an iceberg only by what is seen above the water. Reincarnation allows the spirit to gain further experience on earth toward self-realization. What is more, it is a grace that makes it possible for souls who depart the earth with a dense or heavy karma to atone and to throw off the burden by trying again, and in the process to liberate the spirit for its homeward ascent. And to incarnate again, and again, for as long as the possibility exists in the conditions of the plane.

The living end

If we have past lives on earth it is obvious, with a few rare exceptions, that we do not remember them. Reincarnation is transmigration of the soul. Such notions arouse dread in many, so much so that voices drop at the mention of death and conversation becomes halted. Death is something we shut into hospitals, hospices and care homes. Sometimes fear of death becomes an obsession, even growing into a private hell. Such fears, which would be illogical if there was nothing to follow death, are extended forms of natural fears, such as that of being suffocated, devoured alive by a wild animal, or standing on the brink of an eerie abyss. Religions have played on and intensified such fears, using them to entice people to follow their creeds, rather perhaps than to lead better lives.

But fear of death is not confined to believers. Those who profess no belief in an afterlife wrestle with a sense of foreboding just as much. The word itself is like a black portentous cloud stretched out on a bright summer's day, spreading a cold breath darkly over blossoms, cramping heart and soul. This paralysis is even more poignant when we unexpectedly lose a family member, relative, neighbour, or friend, especially if unprepared we witness the severance of human life amidst our daily routine. Something choking or oppressive overcomes us, however aware we are that death is the inevitable end of life on earth, a perfectly natural and inescapable occurrence. Only when death mows down thousands and millions of people in wars, disasters or epidemics does our paralysis, strange to say, seem to lift. As the experience of wars demonstrates, large numbers of deaths around us make us more stolid, more serene and calmer. We have come to expect and to accept such manner of death. But in 'normal' times something frightening and provocative of frustrated uncertainty adheres not only to the fact but to the very word.

Our life is full of uncertainties, many of which give rise to anxieties and even chronic neuroses. We strive a great deal to lead a risk-averse life. Death, however, is the final uncertainty. When will it come, and what if anything will happen after? Push it out of mind as we may, for most of our lives we are waiting for it; it seems to be a catastrophic event, rather more important than waiting for the result of a job interview or a letter from the beloved, where one could, just conceivably, do something to alleviate the tension. After all there are other jobs, other potential beloveds, but are there other lives?

To ask the question about death really being an end, an irrevocable oblivion, is to invite a thousand contradictory answers in a struggle to comprehend our own mortality. These are not just 'yes' or 'no' answers – there is also the when, why, how and under what conditions that another existence might be conceivable. The safe knowledge is at an end and we move into realms of cloudy speculation. The fact that our consciousness survives after our body suggests that there is an afterlife. Our consciousness lives beyond the body. The body is an implement that consciousness uses. When the body dies, the soul remains. Death is not an end, but a transition from one state to another. If there is no afterlife, as perhaps the majority of people in the western world at least profess to believe, then terror of death is unnecessary. We will accommodate it, just as eventually we accommodated the loss of that job or that beloved. Indeed, our worries should be at an end, since there is nothing to fear of the beyond in time or space, no other life in which to chase shadows. On the other hand, no more cakes and ale, no further scope for ambition, conquest and recognition by our peers. All the same, believing in a life hereafter ought to hold no risk and nothing will be lost. Or will it? Perhaps it is resignation and a deep sense of loss rather than fear that is entrenched in our minds when we think of death. Then survival at any price – as death – is turned into an enemy of life?

Some try to connect with the beyond to obtain assurances that the hereafter actually exists. But somebody who is sure that life with its possibilities continues does not need such continual reassurance, because he already has conviction.

Most people, most non-psychotic or no more than mildly neurotic people, are afraid of death because they are ignorant or unsure of what will happen after. They remain, even at the point of death, preoccupied with the earthly that they will leave behind, rather than reflecting on the path they must inevitably travel beyond the earth. If they could but be persuaded of the certainty that life goes on and that to die is to be permitted to continue striving towards the true home of the human spirit, the ground for fear would cease to exist.

Death on earth is but the expulsion of the human soul from the gross material life to a completely different plane, toward ethereal life. It is changing our habitation from here to the beyond. However, there is no division between this world and the beyond. An unbroken radiation unites all that exists in the entire Creation into the one great reality that is draped into the mantle of the Creator. At earthly death, the spirit has simply shed its outermost covering, the physical body together with its intellect, and may step forth in the ethereal body in the beyond. Death on earth is birth into the beyond, and birth on earth is transition from the beyond. Spirit is our origin, our core and final goal. When all cloaks are discarded upon the great ascent above the material borders, we ourselves are revealed as we truly are – spirit.

Death operates in matter, which swings in a cycle of becoming and perishing, evolution and dissolution, growth and disintegration. We know that matter at the molecular level can neither be created nor destroyed – only changed to another form. What is represented at the molecular level is true at the spiritual level. But the human spirit, a totally different species from matter, is subject to development rather than to mere change, not to the disintegration that leads to the transformation of fresh union. Our spirit steps out of the physical cloak upon departure from the earthly and appears in an otherworldly cloak, often the ethereal body. The recognition of the illusion of death is the beginning of wisdom.

Our spirit should on no account be confused with the body. When the physical body inevitably disintegrates, after the animating life-force has gone, its elements are released to fresh union. Thus death, seen from the earthly perspective as an end, is but the transition of the spirit to another sphere. It is not a call to immobility or a standstill in the midst of life. Rather it is a deeply human process, a companionship at the closing of a fulfilled life. Death is a portal to otherworldly experience, in the same way that every existence on earth, when the soul, in a transition from the beyond, is able to incarnate in a new body, is an opportunity to advance in spirit so that one day we can leave more mature than when we arrived.

Means to heal and curse

We experience our passage beyond the earth ranging from the luminously radiant to the most dark and gloomily ugly, depending on the nature of our inner life, the direction of our striving. In plainer terms – and stripping away the metaphors – the soul of a person, who while on earth steered clear of all recognition of life after death and refused to accept the transcendence of the ethereal, must be purged of its earthbound obsession if it is ever to awaken to ethereal life. Fettered by the narrow comprehension of the intellect on earth, its blind and deaf attitude toward transcendence has blocked the development of its ethereal faculties of perception. At death, it not only finds that it has not ceased to exist, but it is blind and deaf ethereally, unable to experience, and for a duration that could last a number of centuries as though in a deep coma. Indolence of spirit on earth, inexertion of the finer perception, and however erudite such a person might have been, has led to retraction from the finer life.

Death, then, is not simply an end, but a living end and a beginning in one. A human soul journeys into the gross material existence for the purpose of experiencing. It makes use of a physical body. The connection with the body gives the opportunity to dwell for a time on earth, which may be long or short. Whether because of feeble old age, illness or accident, the body is no longer viable and its radiation connection to the soul is loosened, the soul detaches itself and returns to the beyond, and we speak of death.

The dying process breaks the so-called silver cord, which is seen by clairvoyants as a shining 'tether' between the soul and the body. On earth, the physical senses mediate experience of the world around us, but death brings severance of the body and sensation is then, for a brief period, mediated by attributes of the astral body. This explains the observations described in near-death experiences of being drawn in the initial separation through a tunnel, of feeling the self aloft away from the body, or hearing what others still very much on earth are saying, while at the same time experiencing transcendence. With fading distance from the physical body, the departed one may meet relatives or friends that are already departed, whom he now 'sees' again. The finality of death comes with severance of the silver cord and loss of the astral body.

Souls of varied experience coexist and interact on earth. Under the protection of the physical body we can boost each other's progress. After shedding the physical body, we are drawn irresistibly toward the ethereal plane of our homogeneity. It may not necessarily be the case that we will be 'united' with our loved ones from our earth-life, but if the bonds are strong and not caused by guilt, the connection will remain even if the persons concerned dwell on different planes. In the case of strong guilt threads, the person belonging on a higher plane will be forced to dwell on a lower plane with his victim until he has helped him up and thus redeemed his guilt.

To achieve a healthy and fulfilling life, the first thing to be rid of is the fear of death and to embrace life's never-ending possibilities. Only then may we calmly expect the hour when the gate opens for us to the next world. Fear and anxiety of our own mortality paralyse and weigh us down. A positive attitude to death, on the other hand, leads to a relaxed, infective cheerfulness. The cheerful, happy man has drawn from the deepest well of knowledge and has learned the ultimate purpose of our life. Such bliss, of course, does not come from hedonistic self-indulgence or unattainable loud and unrestrained pleasures, but from recognition, a kind of sacred seriousness and joy that comprehends the heights and depths of existence. In this alone the upright, merry life-affirming deed is rooted.

If we have done good and happy works, good and happy conditions await us; if we have sowed wrongly and badly, we will be earth-bound and will need to redeem ourselves. Thus we come anew to earth to gain strength, only very rarely remembering our past existences. None of us are here for the first time – we have passed at least once already through the gate of death. Death educates us – in the primary sense of the word, of 'drawing out' the truth. It 'jars' our alertness, so that we take stock and gain knowledge about our life.

Death is thus not so much a 'good' or a 'bad' thing as an essential and necessary passage. He who seeks to pass through that gate and improve his lot by taking his own life is sorely deluded, for inevitably he will return. By throwing his life away he has thrown away the chance to develop his potential, and in the process to influence others to tread the right path. He has diminished not only his own existence but the sum total of goodness and right thinking in the world. Such damage to the fabric of existence may take considerable earthly time to repair. Then there are situations in which someone takes his or her own life, like the mountaineer who, after a fall, cuts himself free of his ropes in order to prevent dragging his companions down with him, or the pilot who manages to steer the death-dive of an uncontrollable aircraft away from a row of homes into an empty field. Technically speaking, these are both acts of suicide, but in reality they are both acts borne out of instinctive consideration for others in a desperate situation.

Earthly existence is a grace, from the first day to the very last day, of being permitted to experience, so that in the struggle with material friction our spirit will acquire vigour for its own ascent. To shorten one's life from fear of living means arbitrarily to shorten the grace of being permitted to mature. It is to die without honour. Whatever reason may be given for desiring death – terminal illness, misery or pain, despair, remorse or hardship, or enslavement, exile or some other daunting aspect of existence – whoever for all that believes that he can escape is deluding him- or herself. Fleeing from the hardship simply means prolonging it, since it accompanies to the beyond because it was not brought to an end here. Similarly, ending one's life to escape earthly justice will not expiate the crime. Death is no solution and offers no escape; it concerns only the body and not the personally developing spirit within. The sufferings of this earth arise from the consequences of erroneous decisions. In the final analysis, we should welcome everything that life's changes present for the sake of upward inner movement.

Embracing the gift of life

Death often seems to come too early. The gift of life seems to be arbitrarily snatched from some who have barely begun life's journey. At times death appears to seize its victims indiscriminately from all age groups, some being taken in the fullness of rich life activity or soon after they are born. Why were they not allowed to complete the great, beautiful and good things they had begun? In wars, natural disasters and accidents, death finds its victims.

'Arbitrariness' is known only to humans. What we call 'arbitrariness' or 'accident' or 'misfortune' is nothing but the working out of a process, the resonance of which reaches planes beyond our earthly existence, in accordance with the eternal cosmic laws.

This may be a disappointing response to an essentially human question, but if we could but see 'the whole picture' – and from time to time everyone is granted a glimpse – we would understand that nothing of the beauty and variety of the natural world exists without purpose or for no purpose beyond perchance survival. And this is not simply 'blind trust', but a fact of life on earth that every scientist will attest to. It may be little consolation to us in our individual quotidian activities and concerns, but we know for a fact that billions of organisms on this planet die every second and billions more are born into existence, even if we do not know why, like so much else that surrounds us.

Apart from the fact that we carry death around with us every day, spent blood and bodily cells being constantly replaced anew, we need not fear death because it is not only an end but also a beginning. The cycle is seen in every plant and animal, from parent to child, from bud to flower to fruit to seed. Is a seed something that manifests the beginning or the end of life? It depends on one's perspective, but surely it is both evidence of existence and potential. What frightens us, what shocks us most, is the sudden ending of a life, especially one only newly begun. But the parents' pain and their mourning for the loss of the intimately loved young life possibly help to sever a load of karma and clear away an ancient burden. It is ultimately a narrative of hope, humanity and renewal. We can never know how much we atone for and purify ourselves through pain, as also through joy, but in as much as we are readily accepting we will gain clarity in our minds. People who are otherwise apart may be drawn together through pain and suffering. And terrible and humbling as it is to us at the time, we all learn much through the death of loved ones, things we learn in no other way.

As for the millions of victims of wars, these are not due to the blind raging of an impersonal fate, but rather evidence of human guilt and human disgrace. Wars are human sacrifices to false gods, not part of the Laws of Creation. Naturally we learn something from wars (or we should do) about ourselves. For example, the wisdom and knowledge of reincarnation placed against wars that result from clinging to prejudices of race, religion, ancient superstitions, or nationality make a travesty of mankind. We have become dealers in death, and are ever more frantic, ever more blind to higher laws.

Apart from war there are countless other ways in which humans rush heedlessly over the bridge of death, plunging to destruction thoughtlessly, wantonly and without due consideration for the gift of life. The urge to compete and to break records results in the premature death of many, for even the smallest everyday matters are links in the chain of guilt through which human beings shorten existence.

Even in the treatment of the body itself, we increasingly see the abuse of food, drink, tobacco, alcohol and drugs that leads to cycles of unhappiness, soul-starvation and depravity. Fresh air, exercise and the pursuit of *mens sana in corpore sano* (a healthy mind and body) has in large part been replaced by the soporific addiction to spectator sports, and other engagements in which mental activity and personal striving are minimal. Indulgence in such comatose activities, largely underwritten by the profit motive, both private and public, leads to unsatisfying lives, as well as consequent boredom, social disruption, vandalism and violence. Ultimately it must, of necessity, stunt growth of the soul.

What we are doing through such indulgence is not merely shortening and stunting individual lives, but shortening the time of grace that is granted to us in order to mature. One can recognize this most patently and with horror in those who have survived to a ripe old age, thus not having departed prematurely, yet are unawakened in spirit. Old age, which should be venerable in phlegmatic contemplation, having acquired experience, for far too many becomes sad and tiring, the spirit spark of eternal life having been allowed to dull.

There should be no misunderstanding that the mundane is just as necessary to the mastering of our task here on earth as the development of the spiritual. The truth of the matter is that we must put the spiritual above the mundane, striving for it and choosing it wherever possible. If the word 'star' had not become soiled by trite use for blatantly unworthy 'celebrities', we might call the way of the spirit the way to the stars.

Whoever stops moving spiritually binds himself to the planes of matter, which will eventually weigh him down and sink his hampered will in inertia. And he who, firmly bound to matter, looks for the deepest fulfilment in the material may then be not truly surprised if he gradually becomes so sluggish that he is never able to free himself. He may drift in this way through many lives, unperturbed and with impunity, but in the end he will sink into eternal night. For he has committed a transgression against the spirit – the most terrible form of human guilt. This is indeed death, the only death from which there is no more waking. Entangled inescapably, he must then go the way of matter, caught in the mills of disintegration that end the cycle of matter. Such a spirit loses all the personal development acquired so far, which is not mature enough to lift it above the cosmic turning point of material disintegration, and reverts back to its original state as an unconscious germ. It is then discharged from matter to return to the origin, having turned out useless for upbuilding in Creation. At some future point it will have to start all over again, but it will have no recollection of its previous existence which has been totally erased.

The dark silence of spirit death is the real death. earthly death is merely a gate – end and beginning in one. The highest bliss of the perfected human spirit is the personally conscious life in the spiritual realm of Creation, called paradise.

Gleams the winter moon

All development requires time. So, too, the human spirit needs time in the friction of material conditions to ripen to a conscious personality. The span of a single earth-life is not sufficient to achieve this far-reaching development. Further to experiencing in the beyond, the human spirit may incarnate in a new body for the purpose of another earthly existence. It can incarnate in the human form alone, not in a stone, plant or animal. And in the cycle of reciprocal action, the intervals may be stretched between sowing and the harvest and reaping of our deeds. Often these are so distant that we cannot remember the connections.

In reincarnation, the reappearing spirit is of necessity older than the physical body, and may well return to earth a number of times to experience. After attaining a certain maturity that locates it in the finer planes above dense matter, it no longer needs to incarnate on earth. It can continue the path of ascent to the spiritual sphere.

The knowledge of repeated earth-lives is part of the cultural fabric of many communities. It explains fate and justice. In antiquity the doctrine of reincarnation, of rebirth through which the love and justice of God first became intelligible to human spirits, was part of early Christianity, until it was expunged in the sixth century purely for political expediency by the Eastern Roman Emperor Justinian who then held sway over parts of the former western empire as well. There was no conflict between Christian belief and reincarnation.

Certain references in the Bible clearly point to accepted tradition of reincarnation in those days. Matthew narrated in the New Testament the recognition by the disciples that John the Baptist was a reincarnation of Elias from the Old Testament. This account was in part corroborated by Luke (9: 1–9). John narrated the account of a man born blind who was cured by Jesus, whereupon the disciples asked if his blindness was caused by his own or his parents' sins (John 9: 1–12). It is an irony that Christianity does not embrace reincarnation since resurrection of the flesh is a main pillar of faith. Reincarnations, however, are more than an artefact of religious ambition; they are solely concerned with the development of the spirit. Moreover, the cycle of reincarnation is not everlasting. Our universe is subject to a cycle of development and disintegration, and we do not have unlimited chances. The present is a time of a cosmic turning point, of an acceleration of events, which presses the urgency of making the most of incarnation.

We have spoken much of heavy and light souls, winged flights towards eternal light, and so on. If these are but metaphors, they are metaphors that are true to all cultures, and all types of human experience over vast tracts of time. They may not, then, be dismissed as 'mere metaphors'. They are persistent in consciousness and even though they may not be seen or tracked on a screen they are as real as atoms or genes. However, it is obvious that reincarnation, an ancient belief and guide to wise men through the ages, is much more than a figure of speech. Language is not reality, but rather an attempt to track it. What we know is not only considerably greater than what we can talk about, but also more than we know we know. For what we know in the deepest sense is what is in possession of our spirit, what it takes in; whereas we can only say in words what we have learned during our brief earthly span.

But if we, or, more precisely, our spirits, have lived on earth before, why do most of us not remember these past visitations? Most people presume they would remember if they had lived before. This presumption, however, tends to gloss over the fact that there are many forms of memory. For a start we have short-term memory, an intermediate memory and a long-term memory. Anthropologists talk too about race memory. Then we have nightly dreams, where something from out of time seems to be communicated to us, usually to be forgotten or distorted by being interfered with by conscious memory on waking. However, there are exceptional but well-documented cases of people who have lived before. They can remember details from their near or remote past clearly, recalling places where their body has not and could never have visited, speaking familiarly of people long dead, and they even communicate and think in ancient ways.

Using short-term memory we are able to recall the immediate past. But memory weakens as we age, so an old person may forget that a saucepan is bubbling on the stove, or he may lose track of time, or forget where he is, or even his own name. But we all forget things, for as the mind wanders to pick up one thread it inevitably drops another. Only what is converted from short-term to long-term memory is available for later recall. This conversion seems to be linked with certain traces held within the brain. Since it takes time, only a limited number of conversions are possible. Furthermore, the process can be disturbed.

Throughout life we continue to build our memories. Researchers figure that memory is encoded by structural changes in the neuronal connections of the brain, stimulated by the floodtide of sensory stimuli. Thus memories are held as patterns of strengthened signals across the junctions between billions of neurons. The brain is continually building up new connections and remodelling its structure. Each neuron undergoes changes, as do its many synapses.

We say and do many things without being consciously aware of them, or our motivation. We do not remember a lot of what we say and do, and the things we think we can recall accurately are in fact often distorted. We weave some of our present thoughts into memories, and wittingly or otherwise we allow our biases to influence our recollections of past happenings, especially as we filter our perspective of the past through the lens of our current views and moods. Not only can our perspectives change through time, but we also tend to see things as *we* are, not necessarily as those *things are*. Though we may swear in a court of law to tell the truth, the whole truth and nothing but the truth, in reality this is not possible, for with our intellect we see through a glass darkly.

Our memory does yield exact recall. Wisps of information floating about at the time of an experience or at the time of recall adhere themselves to memories. No two individuals' recollections of the same event will tell exactly the same story. Every viewpoint, even that of the camera, is subject to bias.

The frontal brain has but limited storage capacity. The cerebrum can become overloaded and cause memory leaks and distortion. Thus the fact that we cannot remember that something happened does not mean that it did not happen. Often it is the intuitive recognition that can integrate the facts to offer the clearer picture. On occasion we feel that we have been in a certain place before, although we cannot remember when or where. Similarly with people who are complete strangers to us we may well find familiarity. We get good 'vibes' – or the reverse – and act accordingly, with caution or openness, feeling our way, as we say. However, no amount of rooting about in the intellectual storehouse for a clarifying context will enable us to find a clue to where, when or how we may have been in this place, or had this experience or met this person. Animals, of course, have this extra sense in their instinct. This attribute serves them even more strongly than intuitive recognition now does us, because our cerebellum has become stunted.

Reports of earthquake disasters, often with immense loss of human life, tend to be accompanied by stories of strange and unusual behaviour among animals. Such behaviour arises from communication between animals and nature beings (that is, forces of nature that have taken on forms such as elves and nixies and so on), that in the past also alerted human beings to an impending natural disaster.

Sleep – death's counterfeit

Many people think that they do not dream, because they cannot remember their dreams. All that needs to be said is that the dream has not found its way into the long-term memory. During the sleep stages of our dream–wake cycle, we go through dream phases in which our short-term memory participates. Dreaming occurs mostly in the sleep phase characterized by rapid eye movements, otherwise known as REM sleep. If the sleeper is awakened during this phase, the dream lingers in the short-term memory and he is conscious of it. If he wakes up completely and wants to remember the dream, it is converted into long-term memory and is accessible to the waking mind. If the sleeper is not awakened, then he cannot remember it. Thus one tends to remember – and later distort – particularly violent or horrific dreams that cause one to wake up. The tranquil and pleasant dreams, those proverbial sweet dreams, do not arouse the sleeper and are thus far less often remembered.

Sleep interrupts the waking mind and day consciousness. Night rest is a time cue for re-tuning the biological rhythm. The frontal brain must rest after sifting the sensory torrent that showers us all the time, and from synthesizing thought, imagination and other expressions of the intellect all day long. Sleep is the mental restorative that repairs the electrochemical circuitry of our brains. Sleep or a state of rest is normally adequate for the task.

In sleep, when there is a running down of the body's activity and metabolism, there is a discernible diminution of its radiations. The brain waves change and the frequencies drop to a very low cycle. The very low metabolism accompanying deep sleep manifests in a drop in the heart rate, deeper breathing, and lowered blood pressure and body temperature. In deep sleep the body is in a state of total rest and relaxation.

This change in the body's radiation, exemplified by the brain waves, loosens the body's union with the soul. And yet in deep sleep a state of inner animation supervenes with outer complete rest. A person in this phase has complete relaxation of muscle tone and will remain in a tranquil zone until the brain waves rise to a frequency just below the threshold of wakefulness. Then dream sleep, that is REM sleep as indicated by rapidly rolling eyes moving jerkily behind shut eyelids, begins to take over. Early morning dreaming, in which one seems to flicker in and out of strange worlds, in which time often seems dilated – with minutes becoming hours or lifetimes – is like dozing-off sleep, threaded through with strange presences. And surely one clue to the puzzle of why we dream is that dreams alert us to the fact that physical existence is but a passing phase!

A state of outward unconscious inactivity, in which the brain waves approach the flatline EEG (electro-encephalogram) pattern seen in death, coexists with inner animation. But the near 'death' state of the body in deep sleep coincides with the spirit's lively movement, for deep sleep is replaced by an upsurge in heartbeat, blood pressure and metabolism and the brain waves approach their normal waking state. Even while the frontal brain remains dormant, the body remains flaccid and atonic. Indeed, one needs to apply vigorous efforts to awaken a sleeper from deep sleep – it can be like trying to waken the dead.

Dreaming that is unfettered by the resting brain results in a loosening of the physical body. It is like a foretaste of death. The spirit is released and can soar, the sleeper feeling as if borne aloft. Space and time become changed, revealing the developed vision of the active, mobile spirit. The realm in which the spirit now floats is vaster, brighter and stranger than imagination can conceive. One is no longer the same person, but an experiencing entity drifting through space and time. If such surreal experiences on waking seem to rouse unnerving feelings of fear and horror, this is partly attributable to the loss of the intellectual barrier we erect in our waking life. We fear being deprived of the cerebral activity that seems essential for our everyday existence, that which we call 'reality'.

Not infrequently the intellect penetrates dreams, and the unquietened resonance of the huge and over-excited frontal brain infiltrates the intuitional brain. The resulting jumble, like a doubly exposed film negative, generates a confusing anarchy of forms. Hence the phenomena of talking in one's sleep, clenching and grinding teeth, frequent tossing and turning, or groaning, and even, in some cases, somnambulism. Such disturbances in sleep can only be caused by the interference of the frontal brain. Animals, although many do dream, are not prone to sleep-walking!

The spirit knows only the language of pictures, while words belong to the developed intellect. Pictures and words may blend to enable an unhindered, pure volition of the spirit to express in earthly life. Words can form a signpost for the intuitive recognition. In the spiritual realms, however, there are no words as here on earth, but only the intuitive language of pictures. That the spirit in its understanding and comprehending is untouched by vocal language is shown by the fact that people of different mother tongues do not understand each other without a common language. But this gulf of communication is readily bridged when people seek to make themselves understood by signs, since these signs depict moving actions, conveying meaning through moving pictures. Pictures and gestures are thus more primitive – in the best sense of the word – than sounds or words.

Reincarnation by its very nature is a difficult concept for the conscious mind to grasp; it is by definition otherworldly, spiritual and transcendental, dealing as it does not only with matter but matter imbued with spirit. It may be likened to double-entry book-keeping of the soul. There are books that deal with the outer life and which we equate with our memory. We can voluntarily and consciously go back to these books. But there are also books that our volition readily opens but which are closed to our day consciousness. These books are far thicker and can only be accessed by the spirit in moments of heightened awareness. Whereas the day books are logical, sequential and syntactical, the books of the spirit are implicit, emotionally charged, visionary and revelatory of luminous purity in the proximity of the divine.

We carry many memories inside us that we are not conscious of. They may not be accessible to our introspection in our normal waking state, but are no less real for that. Hypnosis and psychoanalysis have attempted to penetrate these unconscious memories, usually with disappointing results, and often with disastrous consequences for the subsequent psychic and emotional balance of the one hypnotized.

Hypnosis has been used to explore unconscious memories and bring them to the waking mind. Investigators have realized the damage they can cause when working with patients, or volunteers in popular venues, either under hypnotic trance or using auto-suggestive means. For in the process, the spirit is bound by an alien will. Only through that which we experience as a free, alert spirit – to which also the world of our dreams belongs – can we be truly advanced. We cannot hand over our soul to someone else in order that he may develop it. Such an act would deny one's free will, the essential birthright of every human soul. Even if some people think they are 'cured' of some addiction such as smoking for example, by way of hypnotism, the craving is only masked and may well reappear in a later incarnation.

A new memory book is laid out for each soul at the beginning of its life on earth, and the soul soon forgets to read the old book from its former life. In fact, the soul not only forgets but is prevented from reading it, so as not to burden its ability to experience.

In exceptional cases the memory book from the former lifetime is not completely closed, which is when a person can remember this or that from a previous time. Such occurrences are not rare and appear especially when only a brief interval has elapsed between embodiments. This can happen when a soul passes on through sudden or unexpected death. It is as if a fruit has fallen from the tree before it is ripe, but is permitted to ripen afterwards. Although such memories are not infrequent, with the past perhaps being jolted by a place or person (that is, déjà vu), for the most part this memory remains unconscious (hence hidden or potential) and only stirs, in an inkling, a feeling of familiarity which may be sympathetic or antipathetic.

No knowledge beyond experience

The fact that we do not remember past lives is perfectly understandable. In the course of a life, from childhood onwards, a certain personality sets in. True, at birth this is not especially marked, but as the years pass the personality unfolds. Even at birth the child is not a blank slate on which experience will inscribe its runes. The Law of Reciprocal Action is active not only within one earth-life but spans the entire existence of a human spirit, including the former lives.

A child is not born innocent, but incarnates its self-formed character and its state of experience. It is embodied with its personally woven web of causes and effects, decisions and consequences – its threads of fate. We give much care these days to the well-being of the newborn, but the body carries something that cannot be seen, weighed or measured; it shields within it a spirit or soul that comes from afar.

The newborn baby at once begins to take in cognitive experiences and his brain feasts on these, and develops at a very rapid rate. Soon he has made these experiences his own. The experiences will gradually become part of him and help him to a psychic ordering of his life. But what he absorbs from the sensory din bombarding him depends on the nature of his personality at birth. The child, in mind and body, is resonant with the vibrations of the spirit that remains sheltered in the early part of life.

Is personality, then, fixed at birth or does it develop along with the body and mind? The answer is that, like the seed, the spirit has potential for growth. It embraces that germ that is eager and hungry for growth. By its nature the essence of spirit is free; it cannot be pinned down, weighed or measured. There is a uniqueness that is there before the embryo, before the procreation, and that is the soul, at its core the spirit, incarnate at birth but destined to rise above the physical.

With growth, subjectivity intervenes. The very young child has no sense of self, having difficulty distinguishing between himself and the world. By the time he has learned speech he will still be saying, 'Johnny likes porridge', or 'He loves his mummy', rather than using the first person pronoun. As the child develops, self-awareness starts to manifest. Shame and embarrassment are complete strangers to the babe in arms and often even to the toddler. During his school years the child becomes more and more a prisoner of his own conceptions and memories, and these harden in the course of his life. In old age the capacity to experience freshly becomes severely restricted because everything new is perceived through the prism of a world-view shaped by memory. Old people often seem to be immured in the past and incapable of receiving vibrations from the world around them.

Spiritually alert senior citizens will recognize this tendency in themselves and thus be able to do something about it, keeping their soul active, refreshing their spirit through contact with the young. They find that the spirit within a person never stops developing, and that they can learn as much from a baby as from the wisest philosopher.

But if an older person is not spiritually alert, he will not notice the ways in which his perception has become dimmed. Then one may say that his soul has no capacity to be impressed, that it dwindles to a mere pinpoint. His earlier expectations of himself and life have narrowed and, frozen in times past, he has become more or less blind to the present.

When the soul eventually leaves the flesh, it is more or less primed for a new kind of growth in conditions beyond the earthly. If it reincarnated again, the vivid memories of a former life would only hamper its development in its present one. The burden of old memories would hinder it from facing the new experiences fresh and unbiased. Like a precocious child, the soul would be convinced that it knew everything and would not be open to fresh experience. In addition, it would be very difficult to have a positive relationship with a person with whom he has issues or unfinished business from a former life and who now crosses his path again in the closing of the cycle. Let us assume for example that a person has committed a gross act of betrayal against another. Upon passing over he recognizes his crime and the wish to make good arises within him. If they both incarnate at about the same time, his earnest volition to make good will lead him to the other person, driven by the desire to be of service to him in some way. He is no longer the same person he was and it may well be that he becomes a most loyal friend to the other. If both knew the details of their past dealings, this would provide a barrier to their reconciliation, even making it impossible.

We live, die and are born again here because we carry our abilities in us still to a large extent in a nascent form. These abilities must first unfold, and it is to this end that life impels us. Not all of our development can be achieved on earth, nor in our brief span of an earth-life. Joined with ethereal experiences in the beyond, more earth-lives will be needed, and many mistakes will be made in each of them. These mistakes must be redeemed through reciprocal action. Thus the cut worm must forgive the plough, the aggressor be pardoned by the one victimized by aggression. All of this takes time, possibly eternities. But we should be sure that no action of ours on earth and no thought is without significance, for there is nothing that we do, think or say that will not incur a reciprocal action: indeed, the fluttering of a butterfly's wing could 'cause' a volcano eons hence!

We are barely conscious of our true self, being as yet unable to distinguish spirit from matter, or core from shell. We each a number of earth-lives behind us, together with the periods in the ethereal world. Birth and death are really the same as we enter and exit the world. It can happen, and is the case in many instances, that souls exit still immature, after wasting so much of their lives in trivial pursuits. They receive another opportunity to put on a new physical cloak for a journey of upbuilding experience. Only in this way can it be said that God is merciful and only in this way can we make any sense of the giddy cycle of our earthly existence, where randomness and fate often seem to us to operate in a void.

Cosmic turning point

Human population on the planet is increasing, in a year-on-year explosion. At the turn of the millennium over six billion people inhabited the earth, with an expansion rate of close to a hundred million a year. In spite of clear fluctuations in the global population, in the past the number of adults has varied very little. Total demographic variation could be put down to the number of young or elderly people, while the middle-aged population remained fairly constant. Similarly, the division between the sexes has been constant at fifty-one percent girls and forty-nine percent boys. In the years immediately following the two world wars of the last century, after a disproportionate number of male deaths, the male birth-rate increased.

The reasons for this global population explosion include increased average lifespan, reduced mortality rates due to improved diet and sanitation, and the development of new medicines, but they fail to satisfy. All these factors are restricted to approximately one third of the globe, to the areas we like to call the developed world. Yet this over-population is not occurring in the West (this includes Japan, too) but in countries that profit less from advances in modern science and technology. Over four and a half billion (close to eighty percent) of the earth's population of nearly seven billion live in very poor conditions, and nearly a billion are close to starvation.

It is hardly sufficient to shrug these facts off; saying that for most people life has always been like that – nasty, brutish and short. The fact is that the gap between rich and poor is wider now than ever before and it is growing more so every day. One might have thought that major breakthroughs in knowledge would eventually lead to global improvements in human conditions. But for most people on this planet the reverse is the case. Where in former times they fought and died by the spear in their hundreds, today they die in their millions, by famine, disease and by powerful weapons sold into their armoury of mass-destruction by others. Poverty and hunger may be attributable to a personal fate, but their effects are multiplied many times by the competitive culture of humanity that often shows little compassion. One might question the nature of man's common sense, as well as his common humanity in a world where starvation is rife yet as much as a quarter of the earth's annual grain harvest is fed to cattle – animals which, for the most part, can live on green fodder.

However, according to the Law of Creation only in giving is there the possibility of receiving anything of enduring value. To *give* selflessly with one's being and nature in relationship with others, to *help* where help is needed, and to *have understanding* for the sufferings and weaknesses of others means to *receive* – that is, to receive something far greater than a tangible reward, namely, a sense of having trodden the simple and true path toward the heights. What a man does for his fellow human beings he does in reality for himself, since according to the cosmic laws everything returns to him without fail, good or evil, in this world or in the hereafter. The reciprocal radiations of his deeds of love, the voluntary consideration and respect for his neighbour that enables the latter to gain self-transcending strength, will quickly uplift him. He will receive new strength through the good deeds and feel lighter in heart as his spirit is borne aloft.

Let us return for a moment to earth and to the population explosion! Despite every attempt to reduce it by birth control methods, the problem of over-population persists. All our best efforts in this direction seem to fail and this is because it is being addressed in the wrong way. It is not a simple matter of biology that is out of control. The difficulties are far more pressing. Attempts to limit population growth using methods like education and birth control clinics, giving rewards for voluntary sterilization, or by withdrawing support from large families, are in the long run bound to court failure. Such state interventions will simply result in bad feeling, laying emphasis as they do on the disparity between the haves and the have-nots. The one-child policy adopted in China, and its strict (one could almost say ruthless) enforcement, has opened up its own chamber of horrors. Second babies are euthanized at birth (although murdered would be the more appropriate term). Baby girls, who somehow escaped abortion, are found washed up on river banks. Because of the quest of Chinese couples for a male heir, male children are kidnapped and sold to other childless couples or to couples who have lost their own son. Quite apart from their moral culpability, these practises will, in the long term, lead to a large disparity between the ratio of males and females, as well as the ratio of the old and the young, which will eventually destabilize both the Chinese economy and their society in equal measure.

From a material standpoint this planet could support and feed even greater numbers than exist at present. Our earth has a huge wealth of food resources – official estimates claim they are adequate to provide for up to twelve billion people. Since the Industrial Revolution galvanized cultural changes in lifestyles over two hundred years ago, humans have been multiplying at an excessive rate. But human ingenuity has enabled us to stay ahead in the race to provide more than adequately for this vast increase in numbers. The fact that some are starving while others are dying of over-indulgence points to a simple issue that involves finding a better method for distributing and accessing those resources. For every physical problem there is always a solution if one looks for it. However, the human being is more than just a body. In reality it is an incarnated non-material spirit. This changes the question, therefore – it is no longer a simple issue of material distribution, but one that relates to the eternal cycles of birth, evolution and decay at all levels of organization of matter and material life.

The cycles happen in time, and time is short. At a certain stage of development the planet earth will no longer be able to accommodate human spirits. We must, by then, have matured to a state where we are no longer in need of experiences in the very dense material domain. There is a finite time allotted for us to develop. For a long time already it has no longer been possible for fresh spirit germs to incarnate in our ripened part of the universe. Development for the old spirits continues via reincarnation. Any thought of migrating to another planet in spaceships if we lose this one can be dismissed as a utopia of the science fiction aficionados, simply because the distances involved are too large and cannot be bridged even at a speed approaching that of light.

Between the physical world, in which we find ourselves at the present time, and the spiritual sphere of our origin, there are numerous planes, and some planes even exist below the earthly. Each plane has its consistent density; the higher, the finer, and the more etherized its nature. In many planes there are spirits whose state of experience is compatible with the conditions prevailing there.

There is an extra factor to the ongoing population increase: the shift in the distribution of human souls between this world and the beyond. One may visualize the whole of material Creation – gross and ethereal matter – like a grand stage partitioned by a moving curtain. During any specific period of the morality drama of human experience in the material planes, the cast is distributed between the front and the back-stage, that is between the visible earthly and the invisible beyond. The earth-lives of the human spirit can be likened to several appearances in the various scenes and acts. Hence a gap exists between what we see and the transcendental reality. Yet in reality there is no division between this world and the beyond. The seemingly impenetrable curtain, fashioned by the over-cultivated intellect, which we have allowed for the most part to override our intuitive capacity, renders the beyond invisible. For, in actual fact, man on earth combines the two worlds in his constitution and can therefore perceive both at the same time if he is inwardly open and thus alert.

At the end of the play the whole cast appears before the final curtain call. Thus we see more cast members than at any time before or in any scene before. A vast cast of characters have wandered through the gross material life. But why so many at this time? Why should this be? One reason is that we are having to incarnate more urgently in the quest to attain our goal of development as the ripened earth approaches its curtain call. It is not only the Judaeo–Christian ethical tradition that has prophecies of resurrection in the Last Judgement, with the idea that evil will be purged from the world. Until recently, reincarnations of human spirits have taken place at what could be called an unhurried pace. This is no longer the case, however, for the advanced over-cultivation of the intellect has led to a cramping of the spirit. More intense cycles of spiritual development are needed in order for us to make good and redeem our misdeeds. In this regard we see all that slumbers in the human soul roused to activity and speedy dispensation in the increased pressure of the living power of the Light that characterizes this cosmic turning point of further evolution in our part of the universe.

The upsurge in the human population on earth is not the result of advances in medicine and hygiene, although obviously these do keep more people alive, nor due to the influx of new spirits. It is due to increased movement in Creation. Our powerlessness in the face of this growth in the world's population is also a recognizable sign of the special character of this development.

Sufficient unto love

A book such as this can do no more than make a gesture towards the light of those eternal verities that belong essentially to the spirit world: truth, beauty and goodness. All of these, as well as their converse, are embodied in human aspirations and ambitions, and in their relationship to the material world. Sages and those who have inklings of the life of the spirit or soul have always attributed these virtues to a deity, who traditionally has been called God. Therefore, in these final words, no apology is needed for using the time-honoured appellations of God, the Creation, His work, and paradise, His kingdom of peace and abode for developed human spirits.

God created His wonderful work of Creation out of His Living Power, which streamed into the void beyond the divine sphere. All that exists forms and shapes its destiny in the one power streaming through and animating Creation, including that of the human being. Although the human spirit is everlasting, it is not created immediately conscious. Impelled in its nature by a gentle urging to become conscious, it leaves the home-world paradise as an unconscious spirit germ to wander through Creation in order to experience, and in the process to develop into a self-conscious personality that can adjust voluntarily to the perfection of the Divine Will that is manifest in the cosmic laws.

Each human being starts out as an unconscious spirit germ. It is his destiny to return to the spiritual sphere. The spirit will return having attained the state of perfection – either that or it will go back as an unconscious spirit germ, stripped of its acquired personal awareness and reverting to its original unconscious state after failing to achieve perfection. Perfection, of course, is never achieved in this earthly plane, but our God-given task is to strive towards that summit. Each person must for himself bear the responsibility for all that he thinks and does. He cannot lean on others, for, although he may seek their counsel, the ultimate responsibility for his activities is his alone.

Truth lies in God's Creation, and recognizing this is the primary duty of humanity. Sadly, despite much earnest endeavour and many seemingly miraculous advances (especially in recent centuries) in the field of science and technology, man has neglected his obligations to God. This is attributable to his pride, his feeling that he – and not God – is the master and that he can 'go it alone'. In reality, man has made rather a mess of God's Creation, debased his own spirit and made something of a mockery of nature's holy plan.

The question may then be asked as to why, since God is all-wise and all-powerful, He does not simply sort out the mess that we have been making. To some this may seem a perfectly fair and legitimate question. However, even to ask it is, in some measure, to attempt to avoid responsibility. The simple fact is that, unlike the beast, we all *know* better than we *do*. In our self-communion, when alone and receptive, and in a calm state of mind in which we are able to receive vibrations from the beyond, we know, in our deepest selves, what the right path is – the way we should have gone, and did not. Such a recognition is the beginning and the end of our salvation on earth.

EPILOGUE

The work of Abd-ru-shin carries a demanding title, namely *In the Light of Truth*. What is truth? Many a great mind has floundered on this simple question, since it is all too obvious that what is 'true' for one is false, or only partially true, for another. We do perceive multiple meanings in the same happening, and experience myriad consequences. For example, two individuals will experience the same colour differently. Truth, thus, may appear to depend on personal experience, on a subjective point of view. It almost becomes, as some have said of religions, a matter of personal preference, a convenience or a matter of choice. One's culture and early conditioning inevitably play their part in what we come to call the truth. Our assumptions and prejudices, those earliest influences of place and person, the very time and culture into which we are born, these become part of our own developing world-view, contributing to our vision of truth. It would seem, then, that we really have no opportunity to judge anything or even to view it without the 'subjective spectacles' of our personal truth.

Is there, then, no such thing as an objective truth? That is, truth that intends to convey its symmetry unchanged through time and space? Abd-ru-shin gives the following answer to this question: 'Truth is the Eternal-Unchangeable! It never changes in its form, but is as it has been eternally and will ever remain, as it is now. It can therefore never be subjected to any development either, because it has been perfect from the very beginning. Truth is real, it is "being"! Only being is true life. The entire universe is "supported" by this Truth!'

A further question may now be prompted by this answer: what is this 'eternal-unchangeable', which remains the same throughout all of time, and which is not subject to any advance or any development? Our material world is marked by a constant change of forms, through things coming into being and dying away. Even our spiritual core is, as all of us experience, subject to a process of maturing and developing. Consequently, it would seem that we shall never be able to find the Truth in the physical world, neither outside ourselves nor within. The answer is that Truth lies in God alone.

But despite the fact that this is hard truth for creatures to digest – that without God there can be no Truth – there is something in Creation for which the description 'eternal-unchangeable' is perfectly fitting: namely, the lawfulness that forms the basis for every movement within and around us: the deep patterns and truths of life, which manifest the unchanging Divine Will. The Laws of Nature and of Creation, described in *In the Light of Truth: The Grail Message* – as the 'Will of God' are indeed unchangeable. And it is for this reason that the description of these Laws in Abd-ru-shin's work takes a central position. Only a view of the world that is built on the basis of verifiable lawfulness can claim to be mediating the Truth, or even to be objectively true for all time.

The Grail Message, however, is in no way merely concerned with a factual, sober and neutral description of nature and its laws; it is above all concerned with the direction and purpose of God's Will. And the fact, therefore, that the Laws of Creation in all things always serve furtherance and upbuilding. Thus we human beings, in accordance with God's Commandments, are essentially part of a process emanating from the Will of God, the Eternal-Unchangeable. Abd-ru-shin's work, therefore, shows us what it means for an individual to know the cosmic laws, the Laws of Creation, and what changes are necessary for him or her to lead a life in accordance with the Will of God. What he lays down are not dogmatic rules of behaviour or any short-term 'doctrine for salvation', but rather an indication to the reader of the right route for his own life 'in the light of truth'. This fundamental work, *In the Light of Truth: The Grail Message* will be a guide for all humanity at its present point of crisis; it shows us why purification is necessary and how it will be beneficial both for us and for the world at large. For thousands of years – a mere blink, it is true, in the eye of eternity – people have been pursuing wrong paths, worshipping false gods created by the intellect and rooted in the material world, when they should have been following paths to spiritual enlightenment. The result has been untold suffering and anguish, all of which we have brought upon ourselves, because nobody forced us to act thus, not even Lucifer. *In the Light of Truth* shows us that the remedy is in our own hands, that we can purge ourselves of the material domination, safely extricate ourselves from the maze formed by the consequences of our decisions, and return to our spiritual home, to paradise, by exerting our divinely inspired power of free will in the knowledge of righteousness.

Abd-ru-shin, whose given name was Oskar Ernst Bernhardt, was born in Saxony in Germany in 1875. Educated as a businessman, his financial position enabled him to travel abroad, and apply himself increasingly to writing. Shortly after his return to Europe, he resided in London. As a result of the outbreak of the First World War, he was interned on the Isle of Man from 1915 to 1919 as an enemy alien. After his release, he began writing the essays that comprise the work *In the Light of Truth: The Grail Message*, finally settling in Austria to complete his work. In 1938 his land and property were seized without compensation by the Nazis, who took objection to his writings because they were diametrically opposed to everything they stood for. Placed under virtual house arrest, he was kept under surveillance by the Gestapo until his death in 1941, which was a direct consequence of the measures taken against him.

Useful resources and reading matter

Websites

www.grailworld.com (GrailWorld magazine)
www.newmpress.com (New Millennium Press)

Literature

Abd-ru-shin. *In the Light of Truth – The Grail Message*. Stuttgart: Stiftung Gralsbotschaft, 2002

Stringer, C., and Andrews, P. *The Complete World of Human Evolution*. London: Thames & Hudson, 2005

Vasey, C. *The Secret of the Blood: Implications for Body, Mind, and Spirit*. Stuttgart: Stiftung Gralsbotschaft, 2010

Fraser, G., Lillestøl, E., and Sellevåg, I. *The Search for Infinity*. London: Mitchell Beazley, 1994

Van Lommel, P. *Consciousness Beyond Life*. New York: Harper Collins, 2010

Bizony, P. *Invisible Worlds: Exploring the Unseen*. London: Weidenfeld & Nicolson, 2004

Popper, K., and Eccles, J.C. *The Self and Its Brain*. Springer International, 1977

Myers, D. *Intuition: Its Powers and Perils*. London: Yale University Press, 2004

Watson, P. *The German Genius: Europe's Third Renaissance, The Second Scientific Revolution and the Twentieth Century*. London: Simon & Schuster, 2010

Darwin, C. *The Origin of Species*, 6th edition. London: Random House, 1994

Index

Abd-ru-shin 90, 96, 207, 279–281
abortion 72, 73, 86, 273
adaptation 39, 40, 42, 44, 49–50, 85, 97, 162
adolescents 180, 199, 200, 201, 202 *see also* young adults
adrenalin (epinephrine) 88, 184
Africa 34, 32–33, 37, 38, 97, 246
afterlife 250, 252 *see also* heaven
alcohol 149, 160, 179, 201, 258
aliveness 208
Alzheimer's disease 68
amino acids 59
animistic substance 163, 164, 198, 199, 202, 230, 242
anthropoids 32, 35, 191
anthropology 27, 29, 31
anthropomorphism 128
antibiotics 191
anxiety 168, 255 *see also* fear; neurosis
apes 32–33, 34, 52, 97, 98, 191
Aryan supremacy 86
astral body 91, 118, 168, 232, 254
atheism 10, 52–54, 122, 220, 234
atomic bomb 69, 81
atoms 104, 109, 110–111, 113, 146, 147
australopithecine 33, 35, 97
autonomic nervous system 167
autosomes 62

bacteria 23–24
balance (sense of) 102
base inclinations 134, 237
Beckett, Samuel 53
behaviour disorders 189
belief 123–124, 125
Bernhardt, Oskar Ernst 90, 281 *see also* Abd-ru-shin
Bhagavad Gita 242
big bang 17, 55, 75–79, 81, 126
biodiversity 85, 97, 245
bipolar disorder 189
birds 50, 66, 162
birth 67, 90, 134, 151, 187, 198, 257, 268, 269
birth rate 272, 274
bisexuality 195, 196
black bodies 105
black hole 82, 243
blank slate 197, 268
blind chance 10, 22, 125
blood composition 180, 182
blood radiation 178–180, 182, 184, 186, 188, 190, 191, 265
blood transfusion 179
blood type 62, 179
blood vessels (fetus) 187
blood volume 179, 185
body–soul union 26–27, 183–185, 190, 231, 265, 266
body temperature 184, 265
Bohr, Neils 108–109
brain development 158
brain disease 154, 189
brain function 26, 31, 66, 88, 144, 152, 161, 165, 181–182, 219, 265
brain size 34, 63, 66, 67, 161, 175–176, 191–193

brain structure 66–68, 156, 170, 262
brainstem 26, 66, 118, 167, 168
breakthroughs 69, 70, 89, 112, 272
breeding (selective) 39, 40, 85, 86
Buddhism 134, 242

capacity for thought 13, 64, 100
carefreeness 180, 181
Carlyle, Thomas 44, 129
central nervous system 167
cerebellum 155, 159, 168, 170, 173, 175, 176, 191, 192, 193, 228
cerebral cortex 28, 66, 154, 155, 159, 167, 168, 170, 173, 175, 191, 193, 263 *see also* brain
chance 1–3, 10, 17, 22, 51, 54, 83, 95, 103, 125 *see also* luck
chaos 11, 95, 223, 235
children 156–158, 180, 185, 190, 193, 197, 199, 202, 268, 269
chimpanzees 32, 34, 37, 38, 60, 64, 97, 98, 191, 273, 274
China 87
choleric temperament 180
Christianity 9, 31, 41, 129, 134, 226, 242, 248, 260, 261, 276
chromosomes 57, 58, 61, 63
chronometric time 218, 219, 220
civilization 244, 245
cloning 68, 69, 70, 71, 72, 73
colour spectrum 106, 108
commonality (ancestral) 35
communication 20, 65, 266–267 *see also* language
compassion 12, 132, 135, 147, 155
competition for resources 40, 44, 85, 98, 120, 203 *see also* struggle
computers 165–166, 226
Comte, Auguste 129
conception 68, 187 *see also* fertilization
conductors (electrical) 112
conscience 33
conscious awareness 100, 119, 159, 169–170, 212 *see also* self-awareness; self-consciousness
consciousness 12, 16, 18, 26, 83, 152, 157–158, 164, 250
Copernicus 45, 53, 130
core (spirit) 118, 144, 178, 185, 204, 228, 269 *see also* spirit core
cosmic dust 84
cosmic laws 22, 90, 114, 143, 144
cosmology 76, 82, 92
creativity 94, 130
Creator 52, 128, 131, 170, 204, 210, 212, 229, 233, 235, 237, 252
cremation 118
Cretaceous period 75
criminality 87–88, 122, 144
crystals 111–112, 140
culture 39, 243–244, 245, 279

dark energy 76, 82
dark matter 76
Darwin Fox, William 47
Darwin, Charles 15, 22, 39, 40, 42–44, 47, 51, 54, 85
Darwinism 23, 41, 43, 46, 99 *see also* evolution
De Broglie, Louis 110
Dead Sea Scrolls 126
death 123, 181, 182, 198, 250–253, 254–258, 265
decision making 30–31, 137–138, 144, 158–159, 247, 249 *see also* freedom of choice
deep breathing 167, 168
Democritus 27
demons 132 *see also* possession
deoxyribonucleic acid *see* DNA
Descent of Man 86
diabetes 59, 182
diet 97, 179, 180, 183–184
dinosaurs 75, 162

disasters (natural) 10, 45, 46, 235, 245, 264
discarnate souls 90, 186, 188
discoveries 94, 95
disintegration 29, 253, 258, 261
divine power 222, 223
divine substantiate 206, 207
Divine Will 130, 208, 212, 277, 280
DNA 15, 20, 32, 37, 43, 55, 58, 60, 69, 72, 74, *see also* genes
Doppler, Christian 78
doubt 52–53, 172
dreams and dreaming 26, 72, 175, 184, 231, 262, 264, 266
drug abuse 149, 160, 179, 201, 258
drug treatments 69, 191

earth-life 134, 136, 138, 139, 169, 194, 233, 247, 260, 268, 271, 275
ecosystems 85, 96
egg cell (ovum) 42, 61, 63, 68, 69
ego 26, 72, 151, 152
Einstein, Albert 14, 66, 108, 124, 248
electromagnetic forces 77, 80, 103, 104, 106, 107, 109, 110, 116, 148
electrons 21, 104, 108–113, 146
Eliot, George 55, 130
embryos 69, 72, 151, 156, 185, 186, 187, 193, 269 *see also* fetus
emotions 154–155, 168
energy 109, 110–111, 115, 116, 146, 222 *see also* quanta
energy constant 107, 108
energy of spirit 141
enlightenment 27, 32, 130, 149, 150, 281
ennoblement 71, 91, 162, 195, 204
environmental change 39, 46
envy 131, 141, 155
eternal unchangeable 279, 280
ethereal phenomena 224, 230, 252, 254, 255, 271, 275
ethnic cleansing 86, 87
ethnic differences *see* race
eugenics 72, 86, 87, 87
euthanasia 89, 185, 273
evil 11, 24, 122, 132, 188, 190
evil spirits 188
evolutionary process 19, 22, 27, 32–33, 34, 38, 41, 84, 98, 163 *see also* Darwinism; natural selection
evolutionary stages (embryonic) 186
extinction 40, 49, 75, 95, 96

family traits 134, 185, 198, 246
fate 135, 137, 144, 248, 249, 268
fatty acids 182
fear 155, 168, 250, 255, 257 *see also* neurosis
feelings 154–155, 168
femininity 195, 196
fertilization 61, 63, 68, 71, 187 *see also* conception
fetus 67, 70, 186, 187, 193 *see also* embryo
feudal systems 128–129
forgiveness 129, 155, 271
fossil evidence 34, 35, 45, 46, 47, 84, 96, 97
free will 18, 24–26, 86, 91, 93, 112, 114, 117, 137, 143, 218, 228, 233, 240, 267, 281
freedom of choice 12, 13, 14, 25, 114, 143, 247
freethinking 127
Freud, Sigmund 15
Fromm, Erich 53

Galileo 45, 130
gender 133, 194, 195, 196, 272–274
gene expression 61, 62, 68

gene structure and function 19, 20, 36, 37, 38, 57, 58, 61, 63, 70
Genesis 44, 163, 232
genetic code 55, 74
genetic diseases 59, 60
genetic variation 37, 38, 39, 43, 60, 62, 64
genocide 136, 246
genome (human) 60, 61, 87
geology 45
glacial periods 36
globalization 176
glucose metabolism 181, 182
goodness 177, 255–256
grace 222, 259
gravity 76, 79, 80, 110, 131, 139–140, 210
Greece (ancient) 41, 128, 150, 205
group soul 72, 163, 202, 230
guilt 137, 240, 258, 255, 259
gut feeling 118, 168

hair colour 37, 62
happiness 143, 178
harmony 24, 244, 245
hatred 153, 155, 168
health 73, 89, 178, 179, 183, 258
heartbeat (fetal) 187
Heaven 27, 49, 117, 142, 150, 205
heaviness of spirit 139–141, 142–143, 168, 261 *see also* lightness of spirit
Heisenberg uncertainty principle 21, 113, 116
Heisenberg, Werner 113
Hell 142
hereafter 252
hereditary mechanisms 20, 60, 90, 174 *see also* family traits
hereditary sin, *see* inherited sin
Herschel, William 54
HMS Beagle 47
hominid evolution 32–33, 35, 57, 66, 97, 98, 164, 191
Homo sapiens 36
homogeneity 90, 98, 131, 255
homosexuality 195–197
Hubble, Edwin 78
human law 121–122, 144
humanity 246, 272
hunger 272, 274
Huxley, Thomas Henry 52
hypnosis 267–268
hypothalamus 167

immune system 191
imperialism 176, 244
impulses 118, 120
In the Light of Truth: The Grail Message 90, 96, 118, 168, 173, 174, 177, 206, 208, 279, 281
in vitro fertilization 71
incarnation (spirit) 90, 147, 150, 171, 186, 190, 193, 247, 250, 260, 268–270
indeterminacy 114, 116
indolence 171, 172, 188, 190, 200, 201, 203, 234
infants 180
influencing others 136, 148–149, 178, 224–225
infrared radiation 106, 108
inheritability of traits 20, 43, 44, 57, 89, 197, 198
inherited sin 173–174, 193
innocence 268
insanity 189
instinct 32, 33, 64, 119, 195, 204, 263
insulin 182
intellect 28–30, 91, 100, 117, 138, 146, 150, 161, 169, 170, 178, 212, 228, 232, 233, 248, 263, 266
intellect (over-cultivated) 275
intelligence 160, 161 *see also* intellect

intuition 13, 28–30, 117, 119, 142, 159, 170, 176, 177, 193, 200, 232
inventiveness 94
Israelites 213

Jehovah 129
Jesus 205–206, 213, 242, 261
joy 168, 178
Judaism 31, 129, 276
justice 25, 121–122, 128, 137, 145, 203, 248
Justinian, Emperor 260

karma 124, 135, 197, 240, 249, 257
Kepler, Johannes 45, 124
knowledge 125, 149, 154, 158, 192
Krishna 242

Lamarck, Jean-Baptiste 42
language 20, 52, 65, 155, 156, 172, 229–230, 238, 261, 266
Lao-Tse 242
larynx 164
lasers 110, 147–148, 149
Last Judgement 124, 226, 276
law giver 95, 121, 122
law of attraction of homogenous species 90, 131–134
law of equilibrium 145
law of motion 145
law of reciprocal action 131, 134–139
law of spiritual gravity 182
laws of creation 258, 273, 280
laws of nature 9, 29, 127, 280
learning 158, 170, 172
life force 126, 127, 222, 233, 253
light (spiritual) 143, 148–149, 227, 241, 276
light energy 53-54, 80, 108, 145–146
lightness of spirit 139, 140 146 *see also* heaviness of spirit
living power 13, 276, 277
long-term memory 262, 264
love 135, 153, 159, 168
Lucifer 240, 281
luck 25, 54, 85, 136 *see also* chance
Lyell, Charles 45

malnourishment 182
manic–depression 189
masculinity 195, 196
meaningfulness 24, 161
meat eating 183–184
media 180, 201
melancholic temperament 180, 200
memory 101, 149, 153, 158, 159, 173, 262, 263, 267, 268, 269, 270
Mendel, Gregor Johann 43, 44
mental disability 190
metabolism (body) 265
metaphor 185, 230, 233, 261
microfossils 84
mind 26–27, 99
mind–body relationship 150, 166, 191
miracles 128, 213
moderation 183
Mohammed 242
monotheism 129
morality 24, 25, 33, 100, 121, 203
mortality 123, 272
Moses 242
mundanity 259
mutations (genetic) 43, 60

natural laws 44, 45, 77, 86, 121
natural selection 40, 42, 44, 48, 49, 51, 99, 100
nature (tampering with) 70
nature (true) 142
Nazism 87, 121, 136, 247, 281

Neanderthals 33
near-death experiences 254
neural plasticity 156
neural processing 165, 167, 174
neurons 67, 153, 154, 158, 173, 174, 191, 193, 262
neurosis 252
neutrons 103, 109
New Testament 261
newborn 156–157, 187, 197, 269
Newton, Isaac 27, 53, 54, 124, 130
Newtonian mechanics 77, 105
nothingness 78, 207
nuclear forces 109, 110, 116
nucleotide bases 59
nutrition 182, 183–184

old age 180, 259, 269–270, 272, 274
Old Testament 129, 234, 261
On the Origin of Species 39
orbits (atomic) 109, 110–111, 113, 146
organ donation 118
orthodox medicine 188–189

pantheism 95
paradise 260, 276, 277, 281
parenting 157, 200, 201
Parkinson's disease 68
particles (energy) 108, 110, 111, 113
passage of time 215, 216–217, 221
past lives 149, 250, 268, 270 *see also* reincarnation
perception 101, 102, 127, 150, 154, 169, 170, 171, 173, 190, 199, 219, 279 *see also* senses
personality 89, 152, 155, 194, 197, 268, 269
personality disorders 179
personification of God 241–242
phlegmatic temperament 180, 181, 259
photons 80, 106, 109, 110, 113, 145, 147
Piltdown Man 126
placenta 186
Planck, Max 105, 107, 108, 124
Planck's constant 107, 108
Planck's quantum of action 113
planes of Creation 208, 210, 225, 242–243, 259, 275
plants 84, 244
Pleistocene period 36
population growth 271–273, 276
possession 188–189
poverty 272
prayer 236–238
pregnancy 68, 186 *see also* fetus
pride 29
primates 34, 35
procreation 195, 196, 202
proteins 58–59, 60, 63, 103, 109, 111, 116
psychiatric disorders 189
psychopaths 88
psychotic behaviour 160
puberty 199
purposefulness 22, 23, 55 *see also* meaningfulness

quanta 80, 106
quantum mechanics 14–15, 21, 77, 80, 82, 103, 107, 108, 116
quantum theory 22, 83
quartz 111
Qur'an 242

race factors 27, 37, 38–39, 243–244, 246, 247
race memory 262
racial prejudice 86, 87, 136, 246
radiation 77, 78, 103–104, 105, 106, 208, 209
real time 218, 219, 220

reality 81, 90, 102, 104, 114, 125, 127, 155, 266, 279
reciprocal action 131, 134–139, 236, 273
red shift 78
reformation 129
reincarnation 138, 151, 198, 247, 249, 260, 267, 268, 270, 274
REM sleep 184, 265
reprehensible deeds 120
reptiles 162
responsibility 135, 166, 200, 277, 278
rest 264–265

sanguine temperament 180, 181
scepticism 10, 160
Schrödinger, Erwin 113
scientific method 13, 19, 27–28
secularism 240
self-awareness 16, 30, 64, 92, 151, 156, 164, 199, 269
self-consciousness 11, 27, 64, 66, 100, 147, 156, 171, 194, 269, 277
self-expression 173
selfhood 65
self-identity 196
self-mutilation 201
self-realization 16, 27, 122
semi-conductors 112
senses 102, 150, 152–153, 154, 167, 173, 219 *see also* perception
serotonin 88
sex chromosomes 61–62, 63
sex education 202
sex ratio 273–274
sexuality 195–197, 201, 202
shame 181, 201, 242
short-term memory 262, 264
silver cord 118, 168, 254
skills 125, 149, 172
skin colour 38–39, 61, 246, 247
slavery 87, 121, 176
sleep 26, 168, 175, 182, 184, 264
sleep walking 266
social class 133–134, 147
social disintegration 29–30
social harmony 11
solar plexus 118, 167, 168, 177
soul 26–27, 72, 118, 142, 156, 160, 163, 180, 187, 198, 229, 230 *see also* incarnation
space—time 79
spectator sports 259
speech 155, 156, 164
speed of light 81
Spencer, Herbert 52
spermatozoa 42, 61, 63
spirit core 16, 71, 72, 90, 139, 146, 176, 198, 233, 247
spirit germ 72, 92, 94, 163, 193, 209, 212, 239, 240, 274, 277
split species 194
stars 79, 80, 81, 84
starvation 274
stem cells 68, 69
stillbirth 73
stroke 68, 154
struggle 23, 35–36, 40, 54, 85, 120, 133, 202, 238–239, 256
subatomic particles 17, 21, 77, 103, 109–111
subconsciousness 72
sugar intake 181, 182
suicide 256
supremacists 86, 136
supreme being 129
survival of the fittest 48, 202–203
swearing 123

tabula rasa 197
technology 69, 93, 125, 226, 277
temperament 180, 200
temperature (body) 184, 265
temperature (energy) 106
terrorists 121

test-tube fertilization 70
The Descent of Man 52
theory of evolution 39, 47, 48, 51
theory of selection 23
theory of transformation 42
thinking process 224
thought control 228
threads of fate 249, 268
thrill-seeking 200
time paradox 214
time perception 214, 215, 218–221
time travel 216–217
time–space continuum 221
tool use 98, 191
torturing 120
transcendence 224, 254
transgression against spirit 259
transmigration of the soul 250
transsexuality 196
true nature 142
truth 122, 125, 263, 277, 279, 280

uncertainty principle 113, 114, 115
unconscious mind 159, 160
universe expansion 17, 77, 78, 82
Unsubstantiate Life 206, 207
upbuilding of Creation 134, 145, 149, 193, 226, 260, 271, 280

vacuum 80, 81, 82
vegetarian diet 183
Victorian society 44, 52
violence 88, 121, 132, 189, 259
vision (visible spectrum) 108
visual perception 152, 154
vitamin D 28
volition 91, 114, 126, 131, 134, 141, 143, 157, 166, 224, 232, 246

wakefulness 264, 265
Wallace, Alfred Russel 15, 22, 42, 99
war 160, 204, 251, 258, 272, 281
Watson and Crick 15
waves (light energy) 104, 106, 108, 109, 110, 111, 113
weak-will 132
weight of spirit 139–140, 141, 142–143, 178, 273
Whitman, Walt 127, 207
Will, evolution of 99
witches 121
womb 186, 266

young adults 180, 188, 199 *see also* adolescents
youth culture 201